SpringerBriefs in Geography

More information about this series at http://www.springer.com/series/10050

Monika Schillat · Marie Jensen
Marisol Vereda · Rodolfo A. Sánchez
Ricardo Roura

Tourism in Antarctica

A Multidisciplinary View of New Activities Carried Out on the White Continent

Springer

Monika Schillat
Universidad Nacional de Tierra del Fuego
Ushuaia
Argentina

Marie Jensen
Universidad Nacional de Tierra del Fuego
Ushuaia
Argentina

Marisol Vereda
Universidad Nacional de Tierra del Fuego
Ushuaia
Argentina

Rodolfo A. Sánchez
Ministerio de Relaciones Exteriores y Culto
Dirección General de Asuntos Antárticos
Buenos Aires
Argentina

Ricardo Roura
Antarctic and Southern Ocean Coalition (ASOC)
Washington, DC
USA

ISSN 2211-4165 ISSN 2211-4173 (electronic)
SpringerBriefs in Geography
ISBN 978-3-319-39912-6 ISBN 978-3-319-39914-0 (eBook)
DOI 10.1007/978-3-319-39914-0

Library of Congress Control Number: 2016940788

Printed on acid-free paper

This Springer imprint is published by Springer Nature
The registered company is Springer International Publishing AG Switzerland

Preface

Tourism in Antarctica has rapidly increased in recent decades. In terms of technical practice and visitor guidance, the management of the tour operators is determining whether tourism has a negative or positive impact on the environment. This book aims to offer new elements for the debate, how co-existence of tourism and environmental protection could best be guaranteed by shedding light on the latest trends and the modus operandi of all parts involved. While some chapters deal with the development of polar tourism in numbers, forms and activities, others will focus on the apparent expansion of new activities carried out in Antarctica and the focus amongst Treaty Parties on the perceived challenges posed by such adventure tourism. The International Association of Antarctica Tour Operators, member organization, which aims to advocate and promote the practice of safe and environmentally responsible travel to the Antarctic, is also part of this study. In this context, special attention will be paid to its strategies dealing with adventure tourism—including both deep field and those additional or new activities taking place from traditional ship or yacht-based platforms. The analysis includes the aspects of risk management and environmental considerations as well as the views on the cultural perspectives on the Southern Continent.

Marisol Vereda examines visitor's representations of Antarctica as a remote destination. Taking into account the number of visitors who travel to the Antarctic Peninsula each summer season, she considers something essential to understand: how they value the area. Her study aims to help us understand the attitudes derived from this tourist practice in terms of a special appraisal of the attributes of Antarctica. This appraisal involves a learning experience, more knowledge about environmental issues and a deeper awareness on the meaning of Antarctica. The destination is associated with special wilderness values, remoteness and exceptional wildlife. These attributes reinforce the idea of the wild, last place on earth with a unique environment and will lead to sensations of remoteness, desolation and being exposed to an otherworldly beauty. Vereda's study shows an exceptional level of satisfaction, which is achieved by the tourists, as their expectations are not only met, but far exceeded by the on-site experience. She concludes that through specific knowledge and direct experience, a deeper state of mind about the meaning of Antarctica has been gained.

Chapter 2, by Monika Schillat, aims to retrace the images of Antarctica as transmitted by the literature. Focusing on the literary production of two centuries, we are invited to revisit science-fiction productions, as well as eco-thrillers and action-adventure novels, as well as non-fantastic narratives, which take place in the everyday experiences on the White Continent and soul-searching in the later twentieth century. Limiting her study to Anglophone readers and travellers, Schillat reconstructs the images conjured in the literature, which might have influenced and motivated the expectations of modern-day travellers. Her article concludes by discussing how the imagery of polar heroes, as presented in the fiction, leads to proposals of extreme adventures and experiences, which might allow the traveller to retrace the steps of their idols from the past.

A different angle is offered in Chap. 3. In conjunction with the rapid growth and diversification of Antarctic tourism since the late 1980s, much emphasis has been put by States Parties to the Antarctic Treaty and academics on the need for regulating tourism activities and monitoring their effects in the Antarctic environment. Environmental non-governmental organizations have also addressed this issue critically. However, issues connected to the supervision of Antarctic tourism activities have been much less addressed by any of these groups. Rodolfo Sánchez and Ricardo Roura address aspects of Antarctic tourism supervision with a focus on shipborne tourism, which is the dominant form of conducting tourist activities in Antarctica. They focus on the way parties to the Antarctic Treaty and the tourism industry supervise the implementation of applicable regulations and conclude by discussing likely mechanisms to enhance supervision of the management of shipborne tourist activities in Antarctica, as well as proposing further lines of research on this issue.

Adventure tourism also poses new challenges for the International Association of Antarctica Tour Operators (IAATO). Monika Schillat focuses on how the IAATO, a member organization that aims to advocate and promote the practice of safe and environmentally responsible travel to Antarctica, has dealt with visitor guidance and the management of the tour operators in the past and discusses how the apparent expansion of new activities carried out in Antarctica are dealt with nowadays. Keeping in mind the aspects of risk management and environmental considerations, Chap. 4 aims to analyze IAATO's strategies towards these new activities.

Chapter 5 deals with the origins and development of Antarctica tourism through Ushuaia as a gateway port. Marie Jensen and Marisol Vereda offer an analysis of seaborne tourism to Antarctica using Ushuaia as a base port from 1958 to 2014. The comparison of different sources enables a revision of the voyages, tourist flows, evolution of seasons and factors that influenced the development of Antarctic tourism and the role of Ushuaia as a gateway city. Likewise, a reference to the "tourist boom" in the 1970s is given, since it is the period when Antarctic shipborne tourism consolidates as a tourist product, being the most active decade before the sustained growth that started in the 1990s. Also, the occurrence of large cruise ships is taken into account. Finally, an integrated analysis per period is provided.

Contents

Chapter 1
Antarctica in the Mind of Visitors: Representations of a Remote Destination

Marisol Vereda

Abstract In recent years, Antarctica has been identified as an important remote destination for tourism, with a diverse offer based on its natural and historical resources, which motivates tourist flows to this continent of the extremes. The mental construct of Antarctica as a tourist destination is specially shaped in the place of origin of visitors by different factors related to the past experience, by word of mouth and external communication as well. This contribution examines visitors' representations of Antarctica as a remote destination. To do so, a structured questionnaire was used during two Austral summer seasons, considering different stages as regards the Antarctic experience (images prior to departure, *on-site* experience and level of satisfaction achieved). Taking into account the number of visitors who go to the Antarctic Peninsula each summer season, it is interesting to consider their profile as well as the representations they have about Antarctica and the type of experience accomplished since the space gains a special meaning and therefore, different tensions take place, which give value to resources and generate attractiveness.

Keywords Tourist representations · Expectations · Satisfaction · Experience · Wilderness

1.1 Introduction

Antarctic seaborne tourism, with a varied offer of themes and activities and operating vessels of different passenger capacity, concentrates visits in the Antarctic Peninsula. Most Antarctic tour operators use Ushuaia, Argentina, as the connecting

M. Vereda (✉)
Instituto de Desarrollo Económico e Innovación,
Universidad Nacional de Tierra del Fuego, Onas 450, 0910 Ushuaia, Argentina
e-mail: mvereda@untdf.edu.ar

M. Schillat et al., *Tourism in Antarctica*,
SpringerBriefs in Geography, DOI 10.1007/978-3-319-39914-0_1

base port, carrying 36,625[1] visitors in the 2014/2015 season. Of this, total, 27,138[2] passengers landed in the Antarctic Peninsula having a direct experience with regard to the different landscape components.

Taking into account the number of visitors who go to the Peninsula each summer season, it is interesting to consider their profile as well as the representations they have about Antarctica and the type of experience expected, in order to know how they value the place.

This contribution examines visitors' representations of Antarctica as a remote destination, considering different stages regarding the Antarctic experience (representations prior to departure, on-site experience and level of satisfaction achieved).

Nowadays, the study of Antarctic visitors' travel experiences becomes more important due to the attitudes derived from this tourist practice in terms of a special appraisal of the attributes of Antarctica. This appraisal involves a learning experience, more knowledge about environmental issues and a deeper awareness on the meaning of Antarctica.

1.2 Tourism as a Social Practice

Every attempt to conceptualize tourism makes its complex character evident. We understand tourism as a socially inscribed practice, clearly modern, being noteworthy the centrality of space since it gets meaning in the social value given to the components of a place capable of generating interest and, therefore, mobilities. That is to say, the reason for taking a trip implies two unavoidable elements: space and time.

In the last decades, the continued growth of tourism has been witnessed, becoming a very common practice. Nonetheless, tourist practices started once spare time had been gained to work time; in modernity it is no longer an empty time, but a time to be optimized (Sue 1982).

The social diffusion of spare time has to do with a social matrix, which clearly distinguishes between work time and leisure time. Modern leisure is defined by three essential characteristics: a material one, available time for the practice of recreational activities; a social one, which refers to the generalization of entertainment for society in general; and the institutional type, where the community is in charge of certain entertainment (Sue 1982). Thus, leisure time constitutes

[1]Information Paper presented by Argentina at the XXXVIII ATCM (Sofia 2015): "Report on Antarctic tourist flows and cruise ships operating in Ushuaia during Austral summer season 2014/2015".

[2]It refers to visitors who travelled on ships carrying fewer than 500 passengers each time. Others sailed Antarctic waters on board vessels that did not land passengers in Antarctica.

merchandise, which has for the dominant system a specific process, production costs and distribution and consuming networks. Such is the case that tourist practices are strongly bound to the split of a sphere of leisure, which has the sphere of work as its counterpart.

Availability of paid holidays, increase of different sectors' acquisitive power, lifestyle changes as well as technological advances, are circumstances that have highly contributed to the development of tourist practices framed in the leisure sphere. Therefore, tourism is seen as one of the products in the commercialization of leisure, being interpreted within the social context where it takes place (Getino 2009).

Accordingly, we define tourism as a social practice typical of modernity (Bertoncello 2002). Albric (2001) refers to social practices as action systems which are socially structured and instituted, taking into account two main factors, on one hand, social, historical and material conditions where they are enrolled and, on the other, the way of appropriation of the individual or the group. He also adds that it is necessary that social practices be appropriated, enrolled in a system of values, beliefs and rules, in order to continue to exist. Besides, Bourdieu's notion of *habitus* makes us think in a system of dispositions in regard to leisure that structures and conditions the ways of doing tourism, being it transformed by the different actions, which are performed (Bourdieu 2007).

In this respect, practices comprise all the movements, frequentations and relationships that are built at different places, that is to say, all the spatial acts that individuals manage in their environments. Practices include the very first learning as far as the most complex social organizations, they repeat and materialize social and spatial interaction, rebuilding legacies and creating novelty (Di Méo and Buléon 2005).

Understanding tourism as a social practice allows us to conceive a fundamental relationship between these practices and the space, considering not only the tourism destination, but also the place of origin of visitors where necessities, motivations and expectations are formed.

1.3 Tourism and Representations

Places and their images are a key factor for the materialization of tourist practices. The tourist demand proceeds from individual and collective perceptions from experiences, which are associated to specific places. Promotion and marketing create dependence in the formation and spreading of attractive and positive images about destinations. In this regard, tourism designs a distinctive map showing us the geography of tourism, as the collective manifestation of perceptions and images that form tourist destinations. However, since perceptions and images are redefined and reshaped according to the changing expectations of people, fashion trends, levels of awareness, mobilities, etc., new tourism geographies are generated (Williams and Lew 2015).

As we have just pointed out, the generation of tourist places is influenced by several factors, performing a tourist dynamic which is based on two dimensions, the material dimension (objective character) and the representational one (subjective character). The former includes all the equipment, services, supplies, employment, etc. The latter refers to the production of narratives (textual and visual) that show how "to read" the place from the perspective of tourism. Representations, which select part of the reality to be shown, are built, giving meaning through social values. Following this idea, we consider that tourism is based on the construction of representations. Therefore, representations refer to images that are generated in the place of residence dependent on cultural matrixes. To some extent, tourism builds the destination as a subordination of the place of origin, the tourist destination is neither the place of residence nor the destination per se, but both associated; this idea makes us understand the tourist place as a very complex one, where several viewpoints, interests and intentions interact.

Even though the concept of representations was first introduced by Durkheim as "collective representations", shared by the same community, in general terms, the perspective of social representations was widely developed by Moscovici (1979), who points out that every representation is formed by socialized figures and expressions, he also adds that a social representation is an organization of images and language because it symbolizes acts and situations which either are or become common. Besides, Jodelet (1986) considers that social representations are presented under different shapes, with different degrees of complexity, images that contain a group of meanings, reference systems that allow us to interpret what is going on and give sense to the unexpected; categories that help classify different circumstances. In sum, it refers to a way of social knowledge, a socially built and shared knowledge, forming ways of practical thought oriented towards communication, comprehension and command of the social environment, both material and ideal.

In this context, we focus our attention on Antarctica as a remote destination for tourism, presenting the way Antarctic visitors value their experience, through the different representations built about the place. Other studies have undertaken this topic, Bauer (2001) presented results on Antarctic tourism taking into account visitors' motivations; Vereda (2004, 2010) analyzed visitors' expectations and satisfaction of the Antarctic voyage in the Antarctic Peninsula area and Maher (2010) developed his research on visitors' experiences in the Ross Sea.

Foreground: Brown Station (Argentina); background: *MV Ushuaia*.
Photo: Author's own.

1.4 Antarctic Visitors' Representations

Given the importance assigned to individuals' representations about the destination, which is the aim of their visit, in this case, Antarctica, we worked on the opinions of Antarctic visitors who went to Antarctica through Ushuaia and who also landed in the Antarctica Peninsula. To do so, a survey was designed, and organized taking into account three phases of the voyage: a priori, in situ and a posteriori. In order to clarify the way different components were considered and how they operate in the voyage as a whole,, we provide a scheme (see Fig. 1.1).

As it was noted in the introduction, the surveys were carried out during the 2012/2013 and 2013/2014 Austral summer seasons,[3] collecting a total of 653 valid forms for the analysis, representing a sampling error of 5 % for a confidence level of 95 %. The socio-demographic characteristics of the respondents are shown in Table 1.1. For most respondents, this voyage was the first (94 %), while 4.8 % had already visited Antarctica once before. When divided by gender, women accounted for 47.7 % and men for 41.5 % of the total number of respondents. The range of age comprised from 16 years old to older than 65. Visitors aged 45–64 represented the

[3]For these seasons 31,994 and 37,164 visitors participated on Antarctic voyages through the port of Ushuaia, respectively. Of these totals, 24,625 and 27,467 went aboard vessels that carried less than 500 passengers being able to make landings at different visit sites (Vereda and Jensen 2014).

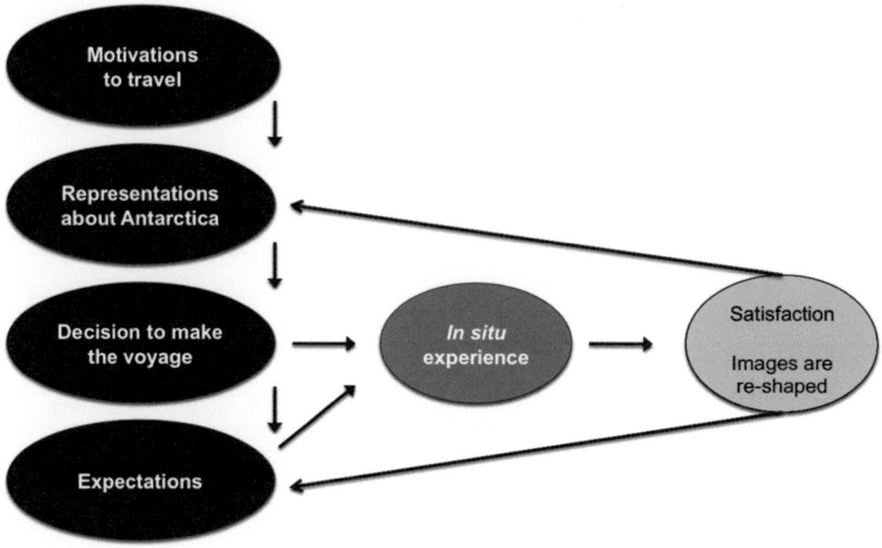

Fig. 1.1 Different stages of the voyage. *Source* Author's own

Table 1.1 Socio-demographic characteristics

Variables	%	Variables	%
Gender		*Education level*	
Female	57.7	Without Studies	0.2
Male	41.5	Primary	0.2
NR/DK	0.8	Secondary	7.4
Age		College/University	91.4
16–24 years	2.3	NR/DK	0.9
25–44 years	24.5	*Occupation*	
45–64 years	44.3	Employed	45.2
65 years and older	28.5	Unemployed	0.5
NR/DK	0.5	Self-employed	16.1
Marital status		Student	2.8
Single	28.8	Housewife	2
Married/living as a couple	58.8	Retired	33.4
Separated/divorced	8.1	NR/DK	0.2
Widow(er)	3.5		
NR/DK	0.8	*Monthly income*	
Place of residence by region		U$S 0–1,700	9
Africa	0.8	U$S 1,701–3,500	25.1
North America	45.5	U$S 3,501 and over	59.3
South America	5.2	NR/DK	6.6
Asia	3.2		
Europe	28.9		
Oceania	15.3		
NR/DK	1.1		

Source Author's own

44.3 % of the sample, followed by the oldest group (28.5 %) and the 25–44 one (24.5 %), the youngest group only reached the 2.3 % of the total. Most of the respondents were married or living in couple (58.8 %), followed by single ones (28.8 %). It is noteworthy that a high percentage of visitors (91.4 %) had a university or college degree. With regard to individual monthly income, the majority of respondents indicated the highest option (59.3 %), followed by the second one (25.1 %), the remaining respondents selected the lowest income. In relation to the place of residence, it is important to mention that most of visitors came from North America (45.5 %), more specifically from the United States (38 %), followed by Europeans (28.9 %), in this case, the most represented nationalities were British (10 %) and German (6,4 %). Visitors from Oceania accounted for 15.3 %, mainly corresponding to Australia (13.3 %). Africa, Asia and South America were the least represented.

Visitors aboard the *MV Ushuaia* contemplate the scenery at Lemaire Channel.
Photo: Author's own.

1.4.1 Antarctic Images Prior to Departure

As we can see in Fig. 1.1, the previous stage of travelling is characterized by the intangibility of the tourist product; the formation of expectations is very much found on the grounds of a flow of information of different sources.

In order to know the perceptions that travellers had of Antarctica before the voyage, they were asked what images came to their minds when they thought of Antarctica as a place to visit. Since it is an open-ended question passengers included

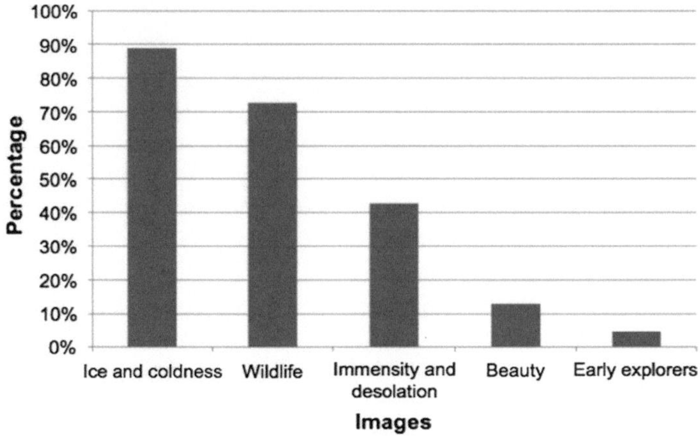

Fig. 1.2 Images in the mind of visitors before taking the voyage. *Source* Author's own

more than one image, so responses could be grouped in different categories. As it can be seen in Fig. 1.2, most respondents pictured in their minds the idea of ice/coldness and wildlife as outstanding images, followed by immensity/desolation. It is noteworthy how this latter category has an important place among the representations of Antarctica, from an affective point of view, but very much associated to the cognitive item ice/coldness of the first category. As regards wildlife, options given by visitors were grouped but it is interesting to point out that penguins in the first place and whales in the second were amply mentioned.

The mental construct of Antarctica as a tourist destination is specially shaped by different sources, which are revised at the place of residence and these work as expectations generating factors (San Martín 2005). These factors involve past experience, by word of mouth and different means of communication coming from commercial and non-commercial sources. In the case of Antarctica, this flow of information may be given through travel brochures, documentary films, early explorers' narratives, photographs, travel book guides, lectures, web sites, etc. In this way, the beginning of an assessment process about the place is encouraged, shaping a mental construct of Antarctica as a destination. Regarding factors that influenced expectations, several ones were identified; results can be seen in Table 1.2. Most factors refer to the past experience of visitors, showing interest in

Table 1.2 Factors that generated expectations

Past experience	%	By word of mouth	%	External communication	%
Interest in natural life	67	Others' recommendation	32	Documentary films	49
Exploration feeling	60			Reading material	30
Antarctic landscapes	59			Early explorers' logs	17
Interest in learning	50			Travel catalogues	15
Interest in environmental issues	36				

Source Author's own

natural life, exploration feeling, Antarctic landscapes, learning and environmental issues. Factors related to external communication include documentary films and general reading material on Antarctica. Practically, all of these items are closely related to sources of information that feed visitors' intention of learning.

Cruising among icebergs.
Photo: Author's own.

Since the place of residence is a determinant issue in the formation of representations, a cross tabulation was made in order to see the relationship of factors and place of residence (Fig. 1.3). The place of residence was grouped in three categories: Europe, Americas (North America and South America) and Rest of the world (Africa, Asia and Oceania).

The above-mentioned factors were quite determinant in the generation of expectations of Antarctic visitors. It is important to point out that expectations are considered as a priori beliefs a person has about the future experience at the destination; anticipation underlies these previous thoughts. In this regard, an open-ended question was asked in order to allow for alternatives to emerge from visitors. Then, six categories could be formed according to the answers (see Fig. 1.4). Is this sense, most respondents agreed on having expectations towards Antarctic landscape, achieving experience, wildlife and learning, whereas fewer answers referred to adventure and exploration and to a lesser extent, quietness.

In order to complete the ideas of representations a priori, a final question was made in an attempt to know what made visitors finally decide to take the Antarctic

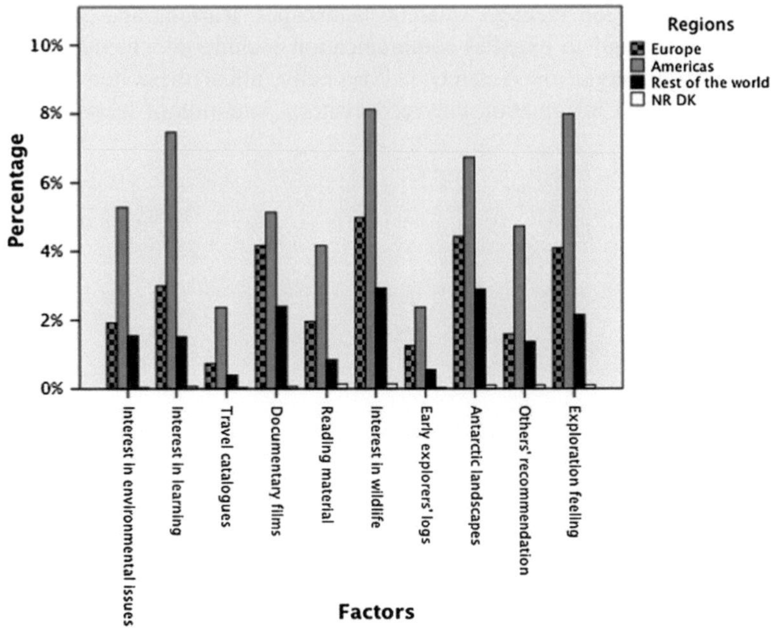

Fig. 1.3 Factors according to place of residence. *Source* Author's own

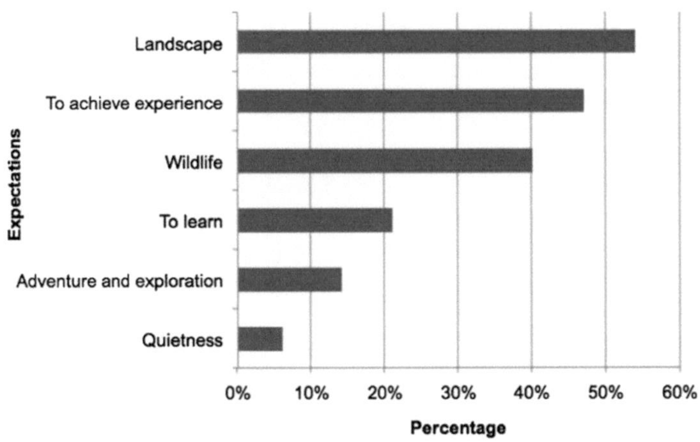

Fig. 1.4 Expectations. *Source* Author's own

voyage. Here, we encounter that decisions were grounded on the fact of achieving experience, partly related to visiting a unique place together with a once-in-a-life experience; nature was also important in the decision-making process. The idea of singular experience accounted for 57 % of answers, of that percentage, 32 % pointed out that visiting the seventh continent was their main decision to take the voyage while the other 25 % agreed on having a once-in-a-life experience. For 27 % of respondents nature was the main item and only a 3 % chose history.

Antarctic visitor capturing photos in the Antarctic Sound.
Photo: Author's own.

1.4.2 Visitors' On-Site Experiences

In order to be able to prepare a list of statements that may allow us to know the appraisal of Antarctic visitors as regards Antarctica during their experience, we worked on two sources of information. On the one hand, previous research work was taken into account (Vereda 2004, 2010) where in the assessment of visitors, different expressions gave value to the natural environment, connected to its aesthetic qualities, the idea of "the wild" was reinforced, and aspects such as world's end, extreme, pristine environment and uniqueness were really important in their representations. Most likely, these outcomes may be related to being far away from everyday industrialized urban places.

On the other hand, aspects that help construct the idea of wilderness[4] from the experience perspective, such as primitiveness (in contrast with modern and technological society), timelessness (possibility to escape the constraints of everyday

[4]It refers to aspects taken from naturalists' narratives, which have been gathered by Borrie and Roggencbuck (2001).

routine, being unaware of the notion of time), solitude (enjoyment of the tranquillity and naturalness of wilderness), oneness (opportunity to re-establish a close relationship with oneself), humility (feeling of insignificance in the presence of wilderness) and care (keeping nature the way it is) were considered. Consequently, 14 statements could be built. Respondents were asked to point out the extent to which they agreed or disagreed with them. In order to do so, a 5-degree Likert scale was used, 1 referred to a strong disagreement whereas 5 indicated a strong agreement. To this end, the arithmetic mean could be obtained for each statement as shown in Table 1.3.

According to the means that were obtained, it is possible to see that all the items reached a high level of agreement, being evident a very reflexive visitors' attitude once the experience was achieved. Item "N" received the highest mark, showing that the voyage turned into an enriching experience, which could not be compared to other trips. Item "K" also became important, since affective attributes were assigned to the destination in relation to a place without interferences. Likewise, items "H", "I" and "A" which refer to the feeling of humility and insignificance at the power of untamed nature, received a high level of agreement; particularly "H" has a special correspondence with the term "awe" which was chosen to qualify the

Table 1.3 Perceptions of travellers through experience

Statements		N	Mean	Standard deviation	Variance
A	I felt totally immersed in pristine and untamed nature	653	*4.53*	0.681	0.464
B	I never cared about what time it was	653	4.13	0.955	0.911
C	I really understood the hardships of early explorers in such an extreme place	653	4.47	0.766	0.587
D	I felt a sense of an earlier and rugged time in history	653	4.15	0.902	0.813
E	I felt a special sense of remoteness from people and cities	653	4.49	0.777	0.603
F	I experienced the simplicity of life	653	3.71	1.160	1.346
G	I felt the heartbeat of Earth	653	3.68	1.150	1.323
H	I was in awe of nature's creation	653	*4.56*	0.788	0.621
I	I felt humbled by all nature around me	653	*4.52*	0.753	0.567
J	Having reached so far South made me feel at the world's end	653	4.14	1.010	1.019
K	I felt the tranquillity and peacefulness of Antarctica	653	*4.56*	0.706	0.499
L	The environment was free of human-made noise	653	4.37	0.890	,793
M	I became aware of the need to take care of fragile environments	653	4.46	0.809	0.654
N	This voyage turned out to be a life enriching experience, opposed to merely a holiday	653	*4.63*	0.682	0.466

Source Author's own

Antarctic landscape or scene. This term "awe" or "awesome" was widely used during the nineteenth century to define sensations related to the big cosmos elements, such as mountains, deserts and oceans. Other important items were "A" and "E" that enhance the value of nature in its original state.

Inspired by Antarctic scenery to draw.
Photo: Author's own.

1.4.3 The a Posteriori Judgement: Visitors' Satisfaction

The process of satisfaction can be defined as a post-experience attitude. It comprises instrumental and expressive attributes; the former refers to the physical products or the means to the tourists' goals, whereas the latter involves responses to scenery, beauty, wildlife and achieving goals for the voyage. (Pearce 2010). Besides, satisfaction is reached as long as the resulting experience adds value to the expectations (Kotler et al. 2005). In this respect, respondents were asked to indicate their level of satisfaction in a 5-degree Likert scale, obtaining very high levels of satisfaction, as can be seen in Fig. 1.4. It is interesting to point out that more than 70 % surpassed their expectations and almost 20 % reached a somewhat expected, in the overall, practically all respondents achieved their expectations (see Fig. 1.5).

As it can be seen in Fig. 1.6, where a cross tabulation between levels of satisfaction and expectations was done, for most visitors who reached the highest

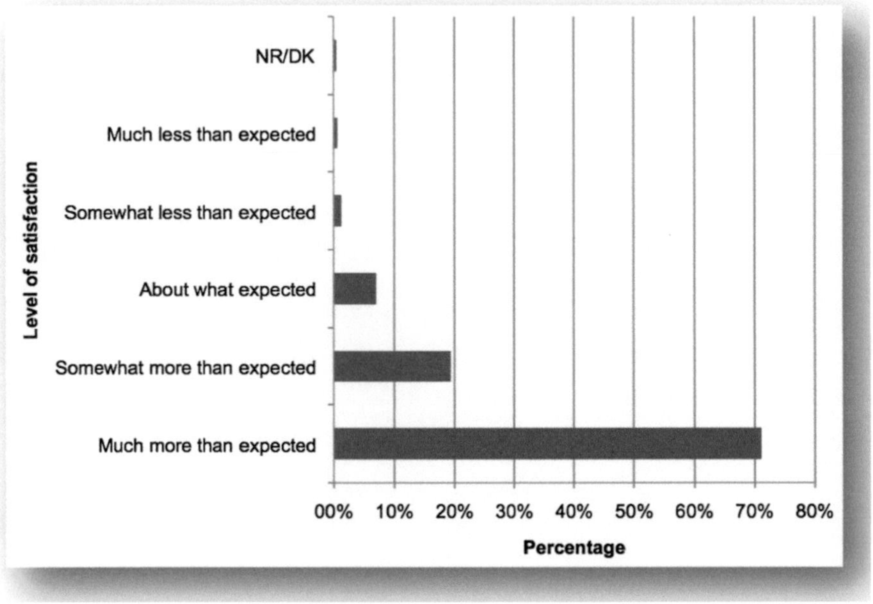

Fig. 1.5 Levels of satisfaction. *Source* Author's own

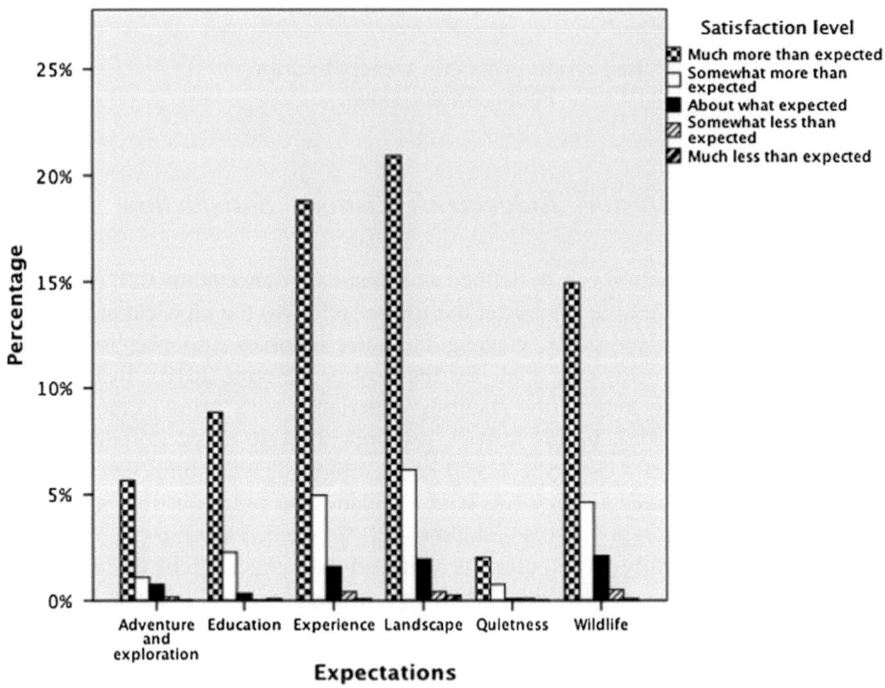

Fig. 1.6 Cross tabulation: expectations-levels of satisfaction. *Source* Author's own

level of satisfaction, their expectations were related to landscape, experience and wildlife.

Also, visitors were asked to indicate why their expectations were either fulfilled or unfulfilled. Among the reasons given for the first case, the beauty of Antarctica was by far mentioned (37 %), followed by the learning experience the voyage was (18 %) and emotion appeared as a non-anticipated expectation (13 %). Remoteness and wildlife sighting were also mentioned, with 11 and 9 %, respectively. Approximately 14 % did not provide any answer.

Concerning the second case, non-fulfilled expectations, since the sum of somewhat less than expected and much less than expected hardly reached 2 %, only 15 % answered this item, including respondents who had referred to satisfaction achieved, but in a way, there were some aspects to consider. The most representative answer with a 6 % alludes to poor sighting of wildlife. It is important to point out that they specially took into account marine wildlife watching, mainly whales and killer whalers. The second reason is related to itineraries that had been changed due to different circumstances (evacuations, weather conditions, ship repairs, etc.), reaching a 5 %. To a lesser extent, noises made by people (2 %) were mentioned and other reasons completed a 2 %.

Visitors admiring wildlife, in this case, gentoo penguins.
Photo: Author's own.

Table 1.4 Main areas of interest in regard to the preferred site

Areas	Places	%
South Orkney Islands (*Islas Orcadas del sur*)	–	–
Elephant Island (*Isla Elefante*) and adjacent islands	Elephant Island (*Isla Elefante*)	3.4
South Shetland Islands (*Islas Shetland del Sur*)	Deception Island (*Isla Decepción*)	7
	Half Moon Island (*Isla Media Luna*)	1.5
	Hannah Point	0.6
	Ardley Island	0.2
	Bailey Head (*Punta Rancho*)	0.2
	Walker Bay	0.2
	Camara Station	0.2
Northeast Antarctic Peninsula	Brown Bluff (*Monte Bardas Coloradas*)	1.7
	Antarctic Sound	1.1
	Weddell Sea	0.9
	Snowhill Island (*Isla Cerro Nevado*)	0.8
	Esperanza Station	0.3
	Paulet Island	0.2
North-West Antarctic Peninsula	–	–
Central-West Antarctic Peninsula	Paradise Harbour	21
	Lemaire Channel	11
	Neko Harbour	6
	Cuverville Island	2.5
	Brown Station	2
	Cierva Cove	1.1
	Danco Island (*Isla Dedo*)	0.3
	Orne Harbour	0.2
	Useful Island (*Isla Útil*)	0.2
	Foyn Harbour	0.2
South-West Antarctic Peninsula	Petermann Island	1.8
	Port Charcot	0.11
	Vernadsky Station	0.5
Antarctic Polar Circle	Stonington Island	0.8
	Fish Islands (*Islote Peces*)	0.6
	The Gullet (*Canal Garganta*)	0.3
	Detaille Island	0.3
	Prospection Point (*Punta Prospección*)	0.2
	Rothera Station	0.2

Source Author's own

Finally, visitors were requested to identify the Antarctic site that they enjoyed the most, for which we obtained a very diverse range of answers. 17 % of respondents said they could not identify a single site since all were relevant. A total of 33 places were mentioned, but Paradise Bay was among the favourite ones with 21 %, followed by the navigation area of Lemaire Channel with 11 % of preferences. Then Deception Island (7 %) and Neko Harbour (6 %) followed in importance.

According to a previous work on visitors' sites (Vereda and Daverio 2011), eight areas of importance for visitors were identified in the Antarctic Peninsula. In Table 1.4, relevance of areas according to the chosen places is shown. Two areas were not considered at all; then Shetland del Sur (South Shetland) Islands and Central-West Antarctic Peninsula were the areas with most selected sites, coinciding with the most visited areas and sites for the last Austral summer seasons, concentrating the favourite sites as well.

Zodiac cruising at Paradise Bay.
Photo: Author's own.

1.5 Final Remarks

Tourist practices originated in the leisure time that has been gained to work time, where men try to find in a determined place the materialization of expectations related to a wilderness, which is perceived as pretty far from the everyday routine, in urban settlements. The expectations may be crystallized through on-site experience. In this respect, to understand tourism as a social practice implies to put it into wider social processes, where it can be interpreted and where it can be interpreted and be made sense of.

From this perspective, the space gains a different meaning and therefore, it enters a dynamic that takes into account both materiality and representational dimensions. From these representations different tensions take place, which give value to resources and generate attractiveness.

In this context, Antarctica represents a singular destination associated to pristine values, remoteness and exceptional wildlife. These attributes reinforce the idea of the wild, last place on earth and unique environment, motivating tourists' visits. Different factors influence these previous representations; in particular, interest in natural life, exploration trips, Antarctic landscapes, interest in learning, documentary films, etc.

Previous images to the Antarctic voyage are basically reinforced from symbols that represent untamed, wild nature from cognitive components, such as those associated to ice—particularly icebergs and glaciers and wildlife, penguins and whales being specially important. The affective components of Antarctica are related to the sensation of immensity, desolation and beauty, which strengthen the idea of a remote and hostile destination.

Visitors' expectations and final decisions to take the Antarctic voyage show a very high interest in achieving an once-in-a-lifetime experience related to unique and extreme sceneries and also to the encounter with marine fauna. These demanding expectations shape an irreplaceable component of the tourist experience, which determines the base of what is expected. The satisfaction process is achieved when the on-site experience is compared to those previous images.

It is worth mentioning the high level of satisfaction achieved since most visitors far exceeded their expectations. Among the reasons for this result, three main aspects emerge: the learning experience together with the confirmation of the Antarctic scenery as a remote and pristine environment and the importance of wildlife. Additionally, emotion plays a very important role with regard to satisfaction, as a feeling that had not been anticipated in the expectations, with a dominance of sensitivity and intuition from a perceptive view. In this respect, through specific knowledge and direct experience, a deeper state of mind about the meaning of Antarctica has been gained.

The favourite Antarctic points of visit are very much related to places that give the idea of desolation, considered as "true Antarctica" and also, areas that can reflect the mental construct of landscape aesthetics.

From the above, we can also conclude that the Antarctic visitor is an experienced traveller who shows a particular interest in Antarctica and who is prepared through reading and other sources of information in order to have a unique experience, being sensitive towards environmental issues as well as demanding with regard to the experience expected.

References

Abric, J. C. (coord.). (2001). *Prácticas sociales y representaciones*. México: Ediciones Coyoacán.

Argentina. (2015). Report on Antarctic tourist flows and cruise ships operating in Ushuaia during the 2014/2015 Austral summer season. *IP 126, XXXVIII ATCM*, Sofia, 01–10 June 2015.

Bauer, T. (2001). *Tourism in the Antarctic: Opportunities, constraints, and future prospects*. New York: The Haworth Press.

Bertoncello, R. (2002). Turismo y territorio. Otras prácticas, otras miradas. *Aportes y transferencias, 6*(2), 29–50.

Borrie, W. T., & Roggenbuck, J. W. (2001). The dynamic, emergent, and multi-phasic nature of on-site wilderness experiences. *Journal of Leisure Research, 33*(2), 202–228.

Bourdieu, P. (2007). *El sentido práctico*. Buenos Aires: Siglo XXI Editores.

Di Méo, G., & Buléon, P. (2005). *L'espace social*. Armand Colin: Lecture géographique des sociétés. París.

Getino, O. (2009). *Turismo. Entre el ocio y el negocio*. Buenos Aires: Ediciones Ciccus.

Jodelet, D. (1986). La representación social: fenómenos, concepto, teoría. In S. Moscovici (Ed.), *Psicología social II* (pp. 469–494). Buenos Aires: Paidós.

Kotler, P., Bowen, J., & Makens, J. (2005). *Marketing para turismo*. Madrid: Pearson Educación.

Maher, P. T. (2010). Footsteps on the ice: Visitor experiences in the Ross Sea Region, Antarctica. *PhD Thesis*. Canterbury: Lincoln University.

Moscovici, S. (1979). *El psicoanálisis, su imagen y su público*. Buenos Aires: Huemul.

Pearce, P. L. (2010). *Tourist behaviour. Themes and conceptual schemes*. Toronto: Channel View Publications.

San Martín, H. (2005). Estudio de la imagen del destino turístico y el proceso global de satisfacción: adopción de un enfoque integrador. *PhD Thesis*. Santander: Universidad de Cantabria.

Sue, R. (1982). *El ocio*. México: Fondo de Cultura Económica.

Vereda, M. (2004) El desarrollo de Ushuaia como puerta de entrada del turismo antártico. Tendencias y competitividad. *Master Thesis*. Sevilla: Universidad Internacional de Andalucía.

Vereda, M. (2010). A study on the expectations of Antarctic visitors towards their trip. Images created about Antarctica and the relationship with Ushuaia (Argentina) as a gateway city. In M. Hall & J. Saarinen (Eds.) *Tourism and change in polar regions* (pp. 236–246). New York: Routledge.

Vereda, M., & Daverio, M. E. (2011). Áreas de interés turístico en la región de la Península Antártica e Islas Orcadas del Sur. Temporada 2010/2011. *Aportes y Transferencias en Turismo,* Año 15, Vol. 1, 83–106.

Vereda, M., & Jensen, M. (2014). La configuración del espacio turístico Tierra del Fuego, Antártida e Islas del Atlántico Sur. El rol de Ushuaia como puerta de entrada a la Antártida. *Final Report Research Project No 898*. Comodoro Rivadavia: Secretaría Ciencia y Técnica UNPSJB.

Williams, S., & Lew, A. A. (2015). *Tourism Geography. Critical understandings of place, space and experience*. New York: Routledge.

Chapter 2
Images of Antarctica as Transmitted by Literature

Monika Schillat

Abstract This chapter intends to retrace the imagination about Antarctica, focusing on the production of two centuries of narration, starting with science-fiction, eco-thrillers and action-adventure novels, than moving on to non-fantastic narratives which take place in the everyday experiences on the White Continent and finally focusing on soul-searching and personal transformation in the later twentieth century. It attempts to reconstruct the images conjured in the literature, which might have influenced and motivated the expectations of modern day travellers. The study is limited to Anglophone readers and travellers. The article concludes by discussing how the imagery of polar heroes of Antarctic Exploration, as presented in fiction, leads to proposals of extreme adventures and experiences, which might allow the traveller to retrace the steps of their idols from the past.

Keywords Antarctic literature · Construction of imaginary space · Tourist expectation · Recreation of epic voyages · Extreme adventure tourism

One might think, that it is almost impossible to imagine a place, which is so very different from what most tourists have experienced before going to Antarctica. But most travellers have a vivid imagination about the White Continent even before leaving home. Visions of Antarctica will most likely be taken from popular films—*March of the Penguins* and *Happy Feet*, as well as documentaries and contemporary novels such as McCaughrean's "*The White Darkness*" (2011), but also literary classics like Poe's "*The Narrative of Arthur Gordon Pym of Nantucket*" (1838) or

M. Schillat (✉)
Instituto de Ciencias Polares, Ambiente y Recursos Naturales,
Universidad Nacional de Tierra del Fuego, Onas 450, 0910 Ushuaia, Argentina
e-mail: mschillat@untdf.edu.ar

M. Schillat et al., *Tourism in Antarctica*,
SpringerBriefs in Geography, DOI 10.1007/978-3-319-39914-0_2

Lovecraft's "*At the mountains of madness*" (1931); from Scott's journals, Frank Hurley's photographs, from travel narratives, including Wheeler's "*Terra incognita. Travels in Antarctica*" (1996) and the much older "*The Worst Journey in the World*" (1922), written by Apsley Cherry-Gerrard. Especially if the traveller is of Anglo-Saxon extraction[1], then he would have grown up reading the stories of Antarctic exploration by British and Australian expeditioners during the "Heroic Age of Antarctic Exploration". Their stories have survived and still work for people very remote from the dead explorers. "One is sitting down somewhere in the warm… and whatever one's attitude, whatever the skepticism one applies to the boyish, adventurous text in one's hands, into one's mind come potent pictures of a place that is definitively elsewhere, so far away in fact that one would call it unimaginable if one were not at the moment imagining it at full force" (Spufford 1997).

But when it comes to retrace the imagination about the Far South, there is more to it, than just the narrated experiences of polar explorers during the Heroic Age of Polar Exploration. The White Continent has turned into a space of imagination, which is the product of two centuries of literary production, starting with science-fiction, eco-thrillers and action-adventure novels, than moving on to non-fantastic narratives which take place in the everyday experiences on the White Continent and finally focusing on soul-searching and personal transformation in the later twentieth century. What most readers do not realize of course is the fact, that this imaginary space has been construed in a complex interaction of different works from the early nineteenth through to the twenty-first century. This is a complex relationship. The act of reading plunges the audience into a web of textual relations, a network of other texts, which altogether create an imaginary space. There are allusions of and dialogues in between different genres even. And this network has produced a certain polar sensibility, which might not necessarily be found in the production about the opposite pole.

This essay is an attempt to reconstruct the images conjured in the literature, which might have influenced and motivated the expectations of modern day travellers. Most Anglophone travellers have been exposed to some of the literary production about Antarctica. But as there is no way of establishing, what has or has not been read by tourists, the selections of works is simply based on their commercial success. This way we can make sure, that at least they did circulate and could have found their way into the hands of an avid reader planning on going south.[2] Analyzing the imagery and engaging thus with this construed space, we

[1]According to the data compiled by the International Association of Antarctica Tour Operators (IAATO) for the Antarctic season 2013–14, 58 % of all tourists visiting the White Continent were of Anglo-Saxon background: North Americans 33 %; Australians 11 %; British 9 % and Canadian 5 %. Information Paper No 44, Report of the International Association of Antarctica Tour Operators 2013–14, ATCM XXXVII, Brazil 2014.

[2]The text selection is based on the commercial success of the works, which can be understood as (a) permanence on bestseller lists; (b) number of editions; (c) quantity of different languages the work has been translated to.

might hopefully get a better understanding towards travels in Antarctica and especially the dreams and expectations of nowadays and future adventure tourists.

Setting the scene in literature tends to integrate prior descriptions of the space and at the same time unfold a complex relation with the imagery present in the representations of this same space. This seems to work surprisingly for literature, which is produced by the so called "armchair travellers" as well as for writers, who have actually been in Antarctica. Both are influenced profoundly by the texts they have come across previously. Even the perception of the writer, in situ is affected by this imagery. Previous texts seem to interact in a creative way in the writer's mind up to the degree where the writer asks himself, "Who has actually invented this, what I am writing?".

The reconstruction of the intertextual structure of novels, travel writing, log books and other narratives could help to understand the imagery about the White Continent, which readers are introduced to. Basically, we are dealing with three strings of narration, each of them at the same time presenting its own discursive rules. The first one might be defined as the scientific discourse, such as presented in expedition reports and log books with its specific style, reduced to facts and details. The second one includes the literature of Proto Science Fiction and Science-Fiction, presenting a mix of the fantastic and scientific as well as horror stories and utopian visions of Antarctica as the place for a better future. The third one is searching for a different way of communicating the special qualities of Antarctica, a world "almost pristine" and of sublime beauty. This romantic way of looking at Antarctica is characteristic for the literature created during the Heroic Age of Antarctic Exploration, but it did by no means end there.

We got our initial clues from the literature studies carried out by Leane at the University of Tasmania. Her analysis of a broad variety of Antarctic narrative stretching over three centuries of literary production has set the tone, when it comes to understanding Antarctic imagery (2011, 2012). Trying to understand a general attitude towards the White Continent, her studies include all kind of different types of literature. Spufford's (1997), "*I May be Some Time*" has been another inspiring source. Focussing on sociological aspects as well as representations of Antarctica, Spufford retraces the almost amorous relationship Anglo-saxons maintain with the White Continent. Trying to understand, why the stories of the heroic age of Antarctic exploration are still appealing to broad audiences, Spufford attempts an archaeology of the myths related to the first Antarctic explorers and gives surprising insights in how these very myths have changed over the years, but never lost their grip. The stories of human endurance at the ends of the earth still do move audiences today. When dealing with discursive lines presented in Proto Science Fiction and Science-Fiction, we consulted several sources, the most promising one being a thesis presented by Wainschenker (2013). The author presents interesting points of view about the representation of Antarctica in movies and literature, where the silence and stillness of Antarctica becomes threatening.

The documents analysed in the present essay are novels, historic and modern travel books, and some few log books. All of them share the same criteria, they have been commercially successful, and most of them have been translated into several European languages, which make it even more likely that Antarctic travellers might have been in contact with them some time before starting their actual voyage. This might have occurred in the form of a theatre play, television series or even a movie, all based on the original literature selected.

As already mentioned, our analysis will keep in mind that a literary production of a geographical space, is usually a product of complex relationships and interactions of a broad variety of texts. The Antarctic landscape, perceived as an enigmatic continent, hostile with a dangerous horizon, can be understood as a construction of uncountable comments. Previous texts are reshaped by new readings and now form a web, which is open to new contributions.

Surprisingly, the first novel ever published about Antarctica was written before the continent was actually discovered, in the year 1820. The author Captain Seaborn, was a firm believer in the "Hollow-Earth-Theory" and this first book belongs hence to the Proto-Science-Fiction line of Antarctic literary production. John Cleves Symmes, a believer in the hollow earth theory, proposed in 1820 an idea about the inside our terrestrial globe, which includes concentric spheres and a hollow interior, which allows people to live inside. The inner world, illuminated by a different sun, could be accessed via both poles. His idea would inspire a long line of science-fiction authors, dealing with Antarctica, such as Poe, "Manuscript in found in a bottle" (1833) and "*The Narrative of Arthur Gordon Pym*" (1838), Verne, who would finish the story, which Poe had left inconclusive, in 1897 under the title "*The sphinx of the ice fields*". In 1931, Lovecraft would revive the sombre ideas presented in both books in his novel "*At the mountains of madness*". Antarctica was still being described as a fantastic place with a challenging mythological geography. In her studies, Leane (2012) identified the main ideas, which would turn into a *leitmotiv* in the literature for almost 200 years. Both poles are considered to be the portals towards the inside of a hollow earth. A great whirlpool drags the ships into the interior of the planet, as presented by Poe in both his works, "*Manuscript found in a bottle*" and "the *Narration of Arthur Gordon Pym*". An unnamed narrator, estranged from his family and country, sets sail as a passenger aboard a cargo ship from Batavia. Some days into the voyage, the ship is first becalmed, then hit by a sand storm with hurricane force that capsizes the ship and sends everyone except the narrator and an old Swede overboard. Driven southward by this magical wind towards the South Pole, the narrator's ship eventually collides with a gigantic black galleon, and only the narrator manages to scramble aboard. The narrator finds outdated maps and useless navigational tools throughout the ship. Also, he finds it to be manned by elderly crewmen who are unable to see him; he steals writing materials from the captain's cabin to keep a journal (the "manuscript" of the title) which he resolves to cast into the sea. This ship too continues to be driven southward, and he notices the crew appears to show signs of hope at the prospect of their destruction as it reaches Antarctica. The ship enters a clearing in the ice where it is caught in a vast whirlpool and begins to sink into the sea.

> Oh, horror upon horror! The ice opens suddenly to the right, and to the left, and we are whirling dizzily, in immense concentric circles, round and round the borders of a gigantic amphitheatre, the summit of whose walls is lost in the darkness and the distance. But little time will be left me to ponder upon my destiny – the circles rapidly grow small – we are plunging madly within the grasp of the whirlpool – and amid a roaring, and bellowing, and shrieking of ocean and of tempest, the ship is quivering. Oh God! And – going down.

Only the message in the bottle will make it back to civilization, the author remains missing. Antarctica turns into a continent of dark secrets. Poe and his followers will repeat their warning messages. The White Continent should not be visited. This last frontier should not be conquered, the dangers lurking there should not be disturbed and awakened. Running alongside and sometimes entwining with the myth of the polar abyss is another set of legends, in which the geographic poles are not marked by the absence but by the presence of other phenomena, such as the large lodestone of medieval legend; the magnetic mountain sitting above a whirlpool in Renaissance maps, as first featured by Gerhard Mercator on his map "*Septentrionalium Terrarum*" in 1595. The lodestone with its mighty power attracts all kinds of metal objects, loosens the nails in the ship's hulls and disintegrates them. Verne used this concept in "*the Sphinx of the ice fields*" (1897). Other geographical notions, such as the idea of a warm polar region hidden by walls of ice drew support from some explorers' accounts and scientific reasoning. In 1823, James Weddell led a British sealing voyage into the far southern latitudes, meeting severe cold weather and seas littered with ice. Pushing south, however, he encountered changed conditions. Whales surrounded the ship, petrels covered the ocean and no ice at all could be seen. The ships reached 74°S in what is now the Weddell Sea, a record southern latitude that held for the next 18 years. Weddell's experience gave credence to the idea of a temperate South Polar sea. The knowledge of the earth's flattening at the poles suggested to some that both the Antarctic and the Arctic might draw warmth from their closer proximity to the earth's core. An open sea at the South Pole appears frequently in nineteenth century and early twentieth-century Antarctic speculative fiction. As other regions of the planet were increasingly explored, this hybrid Antarctica—part fact and part mythology, ice-bound but temperate—continued to flourish as a setting for novels and short stories. Poe and Verne in "*Manuscript found in a bottle*" (1833), "*The Narration of Arthur Gordon Pym of Nantucket*" (1838) and the "*the Sphinx of the Ice Fields*" were both inspired by his observations.

Another captain who inspired writers of their time was the adventurer Morrell, who published his experiences in "*Narratives of Four Voyages to the South Sea, ... and Antarctic Ocean*", (Morrell 1832). On his fourth voyage (1829–31), Morrell was the first to disembark on Bouvet Island, then went to the South Shetland and South Sandwich Islands, where he would neither find fire wood nor encounter any wildlife and barely managed to survive. Penetrating the Weddell Sea, according to his account, he reached the latitude of 70°S and Morrell declared to have passed the South Polar Circle several times. But as the rest of details provided are quite fantastic—he describes cities in the ice, which he could make out in a distance and establishes his position several times quite mistakenly, a 124 miles inland

(200 km). Most readers did not take his accounts seriously. This however did not stop writers such as Poe and Verne to draw heavily from his experiences. Another real-life account, which would influence them, was Reynolds "*Address on the Subject of a Surveying and Exploring Expedition to the Pacific Ocean and the South Seas*" (1836). The first one was based on the 16th chapter of "*The Adventures of Arthur Gordon Pym*" on these publications, whereas the second referenced Morrell and Reynolds throughout his work "*A voyage to Antarctica*".

Sea exploration was a popular literature genre at the time, but Poe was also influenced by Daniel Defoe's "*Robinson Crusoe*" (1719) and Taylor Coleridge's poem "*The Rime of the Ancient Mariner*" (1798), who himself had been influenced by Captain James Cook's voyages of discovery in the years 1768–1779. And as already mentioned, the chasms opening up in the sea at the end of the novel were inspired by the popular Hollow Earth theory of Poe's day. Pym's voyage covers similar geographical and psychic territory. The stowaway protagonist experiences a series of grim events—claustrophobic imprisonment, mutiny, an encounter with a ship full of corpses, cannibalism, shipwreck and a narrow escape from a treacherous Antarctic tribe—before running up against the limits of the world and his sanity. In the last few pages, his boat is pulled towards a terrifying cataract pouring into the earth from the heavens. Before it, its pathway is blocked by a giant white human figure. The novel remains inconclusive. A note of a so called "editor" towards the end leaves room for speculation: "The loss of two or three final chapters (for there were but two or three) is the more deeply to be regretted, as it cannot be doubted they contained matter relative to the Pole itself, or at least to regions in its very near proximity; and as, too, the statements of the author in relation to these regions may shortly be verified or contradicted by means of the governmental expedition now preparing for the Southern Ocean."

Poe and Coleridge turned the White Continent into a gothic locale, with their dark romantic novels. The region had set in train the ghastly experiences of Coleridge's "*ancient mariner*", produced the culminating horrors of Poe's "*Narrative of Arthur Gordon Pym*" and spawned the hideous amorphous aliens of Lovecraft's "*At the Mountains of Madness*" (1931). The narrative centres on an Antarctic scientific expedition which uncovers preserved alien life-forms. These creatures are the "Old Ones"—"star-headed beings" who originally came to earth in its very early history. These come back to life, when the scientists start to dissect them and in retaliation those slaughter some of the scientific team. When journeying into nearby caves, the narrator and his companions are attacked by yet another life-form, the "shoggoths", violent and shapeless beings. The gothic novel exploits US Antarctic explorer Richard Byrd's language describing a land "beyond the pole" and the many "lost race" fantasies set in the Antarctic to invent a scientific expedition that discovers the hideous true creators of the human race dormant under the ice. Ignorantly penetrating the polar abyss, the scientific team is ejected from the ice itself by a volcanic eruption. Lovecraft's apocalypse through the extra scientific discovery of what is the hopelessly inferior and belated position of humans and scientific knowing incarnates the problems of hard limits as one of repressed or unknown origins as well. The narrator insists in warning us, that "It is absolutely necessary, for the peace and safety of mankind, that some of earth's dark, dead corners and unplumbed depths be let alone; lest sleeping abnormalities wake to resurgent life, and blasphemously surviving nightmares squirm and splash out of their black lairs to newer and wider conquests" (Lovecraft 1931).

The idea of disturbing dormant alien life-forms in Antarctica is taken up again a few years later by Campbell in his short story "*Who Goes There*" (1938). In the science-fiction novella, a group of scientific researchers, isolated in Antarctica by the nearly-ended winter, discover an alien spaceship buried in the ice, where it crashed twenty million years before. They try to thaw the inside of the spacecraft with a thermite charge, but end up accidentally destroying it. However, they do recover the alien pilot from the ancient ice, which the researchers believe was searching for heat when it was frozen. Thawing revives the alien, a being which can assume the shape, memories and personality of any living thing it devours.

The novella has been adapted four times as a motion picture: the first in 1951 as "*The Thing from Another World*"; the second in 1972 as "*Horror Express*"; the third in 1982 as "*The Thing*" directed by John Carpenter; and most recently as a prequel to the Carpenter version, also titled "*The Thing*", released in 2011.

> Even ignoring Poe's considerable contribution to Antarctica's literary heritage, the continent's qualifications as a gothic setting are manifold. As a wilderness—and the most extensive and far-flung of wildernesses—it provides a site remote from civilization, on the edge of established social conventions, …As a sublime landscape, it brings the rational mind up against its limits. As a literal underworld, it suggests the monstrous, the infernal, the Satanic. Polar mythological concern with fearful, dark spaces. (Leane 2012)

Antarctica is a more than fitting scenario for horror films. The South Pole is simply put, more remote than the North Pole from the inhabited world. "The Western worldview in which the Arctic rests on the top of the planet and the Antarctic clings, spider-like, to its bottom brings an asymmetry to polar psychotopography" (Leane 2012) This means, that the metaphorical southern journey is not simply a journey inwards but also downwards, a journey that penetrates the darkest, deepest regions of the unconscious. To Antarctica's remoteness and its negative polarity can be added another factor central to its unique combination of spatial qualities—the ice itself. As land depressed under the weight of kilometres of ice, Antarctica is a continent of buried secrets. It is no coincidence that many far southern horror stories involve a journey not only to the ice but under it, through fissures, crevasses and tunnels to subterranean caverns. There is something that lies there, something hostile and deeply threatening to human reason. Wainschenker (2013) relates this necessity to keep Antarctica isolated to the notion of timelessness. "Not only is Antarctica isolated from the rest of the world, but also from time itself. Time seems to freeze here". Forever unchanging, Antarctica seems to be untouched by time, an idea, which still is present in modern day novels, as we will see later.

Following the line of science-fiction works, we consulted the most successful modern day novels of the genre. Robinson, "*Antarktos Rising*" (2009), Batchelor, "*The Birth of the People's Republic of Antarctica*" (1981) and Robinson,

"*Antarctica*" (1997), conjure up an apocalyptic world, which races to claim a new continent, Antarctica, as the only possibility to survive. In the novel, "*Antarktos Rising*" (2009), they will have to face the fact, that Antarctica is already taken. A phenomenon known as crustal displacement shifts the Earth's crust, repositioning continents and causing countless deaths. In the wake of the global catastrophe, the world struggles to take care of its displaced billions. But Antarctica, freshly thawed and blooming, has emerged as a new hope. Rather than wage a world war no nation can endure, the leading nations devise a competition, a race to the centre of Antarctica, with the three victors dividing the continent. But the dangers awaiting the winning team are far worse than feared; beyond the sour history of a torn family, beyond the nefarious intentions of their human enemies, beyond the ancient creatures reborn through anhydrobiosis—there are the Nephilim, descendants of extra-terrestrials and humans. Morris in "*The Icemen*" (1988) and Botaya in "*Antártida 1947*" (2010), convert the Antarctic into a space, where German Nazis have found their sanctuary in a secret military base. Once again the danger emerges from the White Continent. Batchelor on the other hand positions himself openly in the tradition of the works of Poe, "*The Narrative of Arthur Gordon Pym*" and Melville's "*Moby-Dick*", recreating an epic adventure under the title "*The Birth of the People's Republic of Antarctica*" (1981), As civilization teeters on collapse and national boundaries are closed with governments doing nothing more than announcing "Trespassers Will Be Shot!", a group of disaffected malcontents who had lived on the fringe even in the good times undertake a voyage to escape a glorious socialism that excludes all but a select few from benefits. What begins as a voyage of salvation rapidly becomes a journey rivalling Dante's descent into Hell. The group—idealistic anarchists who are joined by extended family members and an egomaniac opportunist in the guise of "the clear thinker," sail from Sweden's west and south, eventually arriving in Antarctica. They find it has become the dumping ground for the perceived riff-raff of the world. It is nothing less than a vast concentration camp administered by presumably earnest charities and aggressively pragmatic governments operating under the notion that liberal and progressive words can mask what is really underway. It is here that Grim Fiddle—the leader of the group—moves into the forefront of a rebellion that is more a response to certain death than political motivation. Robinson returns to the idea of a utopic society on the White Continent in her novel "*Antarctica*" (1997) and at the same time her text dialogues with Roald Amundsen's and Robert Falcon Scott's writings. Robinson takes us to a harsh, alien landscape covered by a sheet of ice two miles deep. A stark and inhospitable place, its landscape poses a challenge to survival; yet its strange, silent beauty has long fascinated scientists and adventurers. Now Antarctica faces an uncertain future. The international treaty that protects the continent is about to dissolve, clearing the way for Antarctica's resources and eerie beauty to be plundered. As politicians and corporations move to determine its fate from half a world away, radical environmentalists carry out a covert campaign of sabotage to reclaim the land. The winner of this critical battle will determine the future for this last great wilderness.

In the twentieth century, new images would be added to those of an impossible and improbable landscape. It is the images of sufferings and heroism of Antarctica's explorers and scientists, in their majority of Anglo-Saxon extraction. Their time of exploration was called the "Heroic Age of Antarctic Exploration" bearing in mind, that a lot of human endurance and suffering was involved. The most important ones in this context being: Robert Falcon Scott's voyage on board the *Discovery*, 1901–04; Ernest Shackleton's on board *Nimrod*, 1907–09; Scott's following voyage on board *Terra Nova*, 1910–13; the Australasian Expedition under the command of Douglas Mawson on board the *Aurora* and Shackleton's Imperial Trans-Antarctic Expedition on board the *Endurance*, 1914–16. From their books, men with frozen beards are staring right back at us. Not even the yellow stains on the photographs and their frames can take away this sensation: they still seem to be amongst us. The prolific work of writers have made sure that their myth will not be forgotten. Considering the historic context of the production of this kind of literature, the generally sportive and competitive attitude of the early Anglo-Saxon explorers and the extraordinary superb landscape they encountered, it is not surprising that they would have chosen a romantic language to describe the sublime White Continent. Astonishing however its perseverance through time is. The "sublime" is kept alive in clichés about this "pristine and peculiar" landscape for centuries without fissures. The significance of Antarctica for humankind derives from the fact, that there "man is confronted with a world, which has developed in his absence" (Mickleburgh 1988). This is how he learns his place in this unique landscape, which not only shows how insignificant he is but also stays indifferent to him.

In this context, Scott reportedly has become the "iconic British hero". The story of Scott's men has taken on a mythic meaning in its 100 years of evolution. The same is true, to a lesser extent, of other Heroic-Era epics: Amundsen's assault on the pole, mythologized in a very different way to Scott's; the "Winter Journey" of Cherry-Garrard, Wilson and Bowers—a quixotic quest in search of penguin eggs that has come close to supplanting the polar journey as the classic Antarctic narrative; the crushing of Shackleton's ship the *Endurance* and the trials and triumphs that followed; Mawson's solo trek—"the ultimate Antarctic saga"—following the shocking deaths of his two companions, Ninnis and Mertz. These are the origin stories of a continent bereft of indigenous inhabitants and corresponding creation myths. Like all origin stories, they are re-told and re-interpreted by each generation, mocked and venerated alike, but never lose their grip on the popular imagination. The stories of the Heroic Era are repeated in many forms: in popular histories; in television documentaries and dramatized mini-series; in expeditions and tourist cruises that promise to follow "in the footsteps of" the early explorers; and in imaginative works. They are re-told with admiration and nostalgia; re-enacted in imitation and homage; re-imagined from new perspectives; re-evaluated in the light of new knowledge about the circumstances; and re-thought from different political viewpoints. As Spufford notes, "Like any successful myth, [Scott's story] provides a skeleton ready to be dressed over and over in the different flesh different decades feel to be appropriate". Literary responses to the polar tragedy date from the time it was first reported in February 1913 and might testify to the British admiration for their polar heroes as well as for their romantic relationship with the cold and the wide Polar Regions. Grace Scott, the explorer's sister, tried to explain her brother's motivation and what had driven him to attempt to conquer the South Pole: "He felt in himself keenly the call of the vast empty spaces; silence; the beauty of untrodden snow; liberty of thought and action; the wonder of the snow and seeming infinitude of its uninhabited regions whose secrets man had not then pierced, and the hoped-for conquest of raging elements" (Spufford 1997). Surprisingly his motivations, do not seem to be very different from those of modern explorers.

The most successful contemporary writers (Diski 1997; Campbell 1992; Wheeler 1996; Keneally 2001; Bainbridge 1993) still do follow the clues their polar heroes left, when they risked everything in this "pure" and "still intact" landscape. A first glimpse at their introductions reveals their venerations of the ferocious icescape of Antarctica as the last frontier on earth. "My name is Morgan Lamont. As I begin at last to tell this story, I am dwelling in a place where few of you who read it will ever have been; it is a harsh place, and a beautiful one" (Arthur 1999). Thus begins the novel "*Antarctic Navigation*", with the narrator introducing herself. Since childhood, Antarctica has been Morgan Lamont's passion. A strong fascination for Robert Falcon Scott turns finally into the necessity to follow his footsteps. Morgan is a woman driven by a wildly heroic obsession… Barring a trip to the continent itself, there may be no better way for her to experience the perilous and endangered majesty of Antarctica. At the same time, she is not driven to conquer, but to understand and preserve it. Other authors establishing a dialogue with their historical heroes, such as Beryl Bainbridge, who gives each expedition member their own distinctive voice to the story of Scott's ill-fated expeditions in "*The Birthday Boys*" (1993), offers a fresh account of the horribly familiar story. Bainbridge evokes an unendurable landscape without and the chilling interior landscapes of damaged souls. Kitto, in the "*Antarctica cookbook*" (1983) manages to merge the tradition of Antarctic Science-Fiction writing with the historical dramas of exploration. His main character succeeds in time travelling and decides to offer a hot beverage to Shackleton's marooned men on Elephant Island in 1916. But the expeditioners decide that this generous offer cannott be real and prefer to think of it as a "mirage", better to be ignored. They prefer to go hungry instead.

Lately the idea that the hostile and dangerous Antarctic Continent should be closed to human visitation for their own good, has also lead to the demand that visitation should be reduced to armchair travelling only: "For the first time since getting to Antarctica, I was afraid. … Mine was a nameless, shapeless fear. The singing, raging happiness inside me—at the vicious beauty of this place—had drained away, and I liked myself better when I was the one person not afraid. At home, I could have shut the book and put it back on the shelf. Now somehow Antarctica had overspilled the binding, overrun the bounds of safety" (McCaughrean 2011). But there is also the idea that it is the rest of the world, which poses threats to Antarctica and its environment. In representation of others, we'd like to mention the German bestseller by Trojanov (2011), "*Eistau*". A glaciologist, who loves his field of work and especially glaciers, despairs when he understands, that he cannot save the glaciers of the Alps from melting away. He signs up to be a lecturer on board a cruise ship to Antarctica instead. On board, he tries to persuade the passengers of the importance to save the Ice of the White Continent, but only encounters indifference and lack of interest. This leads him to take desperate measures. The concern about the Antarctic environment often turns into the demand to ban tourism from the White Continent, but this demand of course can have a contrary effect in the future traveller. Just like tourists hurrying their voyage to Venice assuming, that it will be buried under the water of the lagoon soon, the traveller, is lured towards Antarctica the more hostile and forbidding it is presented. Operators do react to this demand by offering extreme adventures on the polar ice. Races over the plateau, skiing expeditions and even traverses on foot to reach the last frontier of our planet, are organized every year with increasing tourist numbers. In the austral summer 2011/2012 alone, 250 tourists were expected to reach South Pole Station on foot or with skis. Their aim was to follow in the footsteps of their heroes: Amundsen and Scott. The continent has become a place of soul-searching and personal transformation in the later twentieth century, which is most likely due to changes in the nature of Antarctic travel from the 1960s. Especially during the last two decades, a new component has been added. More and more travellers are looking for extreme adventures on the Ice, asking to be challenged and brought to their limits, following in the footsteps of polar explorers from previous centuries.[3]

The sense of Antarctica as a place apart means that it could also be considered a time apart, where the stories freeze, as well as men and animals. They remain forever suspended in the ice. Time stands still in a frozen world. The main character of the "*The White Darkness*", dialogues with an imaginary Titus Oates, who keeps her safe and guides her through a severe whiteout: "It might be the twenty-first century it might be 1912. Minutes or whole years might be passing, but he is carrying Time, too, inside his useless, frost-bitten fists" (McCaughrean 2011). This timeless quality of Antarctica leads to the illusion that one could actually retrace the steps of Antarctic heroes and repeat their hardships and endurance. Antarctica has been explored and mapped. Yet the minute the modern explorer steps out of the plane or leaves his base camp, regardless of all his hi-tech equipment, he is in exactly the same Antarctica that Scott and Shackleton travelled in. It is remote and it is hostile. Making the slightest mistake can put one's life at risk. It is an unforgiving place. Colder than cold, bleak, a vast wasteland of iciness, its deadliness stretches for thousands of miles.

An ever growing number of touristic expeditions try to recreate epic voyages of the past, such as Douglas Mawson's expeditions on the White Continent or Ernest Shackleton's open boat voyage from Elephant Island to South Georgia. The latter even includes the traverse of the densely glaciated interior of the island.

In 2013, new ingredients were added to the commemorative expeditions. All of a sudden it was not enough to use the replica of Shackleton's open boat the "*James Caird*" to re-enact the crossing from Elephant Island to South Georgia. In addition,

[3]Information Paper No 78 Adventure Tourism: Activities undertaken by IAATO Members, ATCM XXXVII, Brazil 2014.

the expeditioners had to use historic clothing and the same inadequate food their heroes consumed a 100 years ago. Jarvis for instance, led an expedition were pemmican was the only food source for the participants, together with a couple of cups of hot chocolate per day (Jarvis 2014). It seems that there might be more extreme voyages being offered by tour operators in the near future, which might replace more traditional expedition cruises focusing on educational and contemplative aspects. Passengers seeking to test their own limits, physical and psychological, in extreme adventures, demand excursions, which do include kayaks, diving, mountain and ice-climbing, as well as overnight stays in tents. To them it seems the only way to measure up to their heroes.

Others—based on the idea implanted by the literature that Antarctica is a place where time itself seems to be frozen—suspect, that there are lost civilizations hidden under the ice. Groups of spiritualists come to Antarctica in search of esoteric entities, which might give them answers to pressing questions about the past and future of our planet or look for the long lost Atlantis. This search for real and unreal sensorial experiences also seems to be triggered by the presence of meteorological phenomena, which are not easy to be understood, such as the reflection of the sun in very cold air. "Quite suddenly the fog changed substance over our heads. In a matter of moments, the fleshy grey mist resolved itself into frozen dew, a precipitation of crystals, a burden of ice particles that fell twinkling out of the air like rice at a wedding, sunlight splitting them in to all the colours of the rainbow. We were bombarded with rainbows falling from an infinite heights, dazzling us with iridescent spears and darts and cataracts of cascading colour… The fog was gone—a magician's cloth deftly whipped off a table of marvels. In the sky, the sun was a hub of dull aluminium spoked with strands of light, and at the end of each spoke—another sun. Cloned suns" (McCaughrean 2011). And just like other travellers, the narrator of this novel is having trouble to accept, that she is not in the presence of the supernatural, when contemplating mirages over the horizon, produced by layers of air with different temperatures. "It's a mirage," said Titus. "Mountains a hundred miles away." But I did not want it to be mountains a hundred miles away. I wanted there to be people, sentries, Martians in a flying palace of a ship; a secret US establishment we had stumbled upon by chance. I wanted it to be Aeolus, brass-walled home of the King-of-Winds, shipwrecked here in the days of myth. I wanted so much for it to be real. In a place were 'real' puts five suns in the sky and slices rainbows into sushi, why shouldn't there be a palace adrift on the Ice?" (McCaughrean 2011).

Dreams of a more primitive nature come to mind. In Antarctica, it seems, the world is still young, almost untouched by human intervention. A certain appetite for the frozen beauty of a world apart is awakened. The hostile landscape with its forbidding horizon poses a challenge at the same time. And the authors insist of sending warning messages. This continent should not be visited. But there it is, this ultimate frontier, which seems to be dangerous and hence is very attractive at the same time. The voyage today, as it was for the early explorers in their days, always was twofold: facing adversities on the outside and engaging in a voyage towards the deepest layers of our own unconsciousness, towards the dark spots, where dreams, fears and old traumas lurk.

The introduction to Diski's novel "*Skating to Antarctica*" (1997) might serve as an example. Her book is based on her personal life experiences and a voyage, which she took on board an expedition cruise ship to the Antarctic Peninsula. The main reason for going south seems to be an inner search, the idea to find herself in the process. Diski projects her interior onto the white spaces of Antarctica, building a mental landscape seemingly free of painful memories, a place possibly able to transform and heal her. "I am not entirely content with the degree of whiteness in my life. My bedroom is white: white walls, icy mirrors, white sheets and pillow-cases, white slatted blinds. It's the best I could do. Some lack of courage—I wouldn't want to be thought extreme—has prevented me from having a white bedstead and side-tables. Opposite my bed, in the very small room, a wall of mirrored cupboards reflects the whiteness back at itself, making it twice the size it thought it was." Diski admits, that it would have been much easier to travel north to the Arctic starting from her home in England. Just like a sexual compulsion, somewhat annoying and inconvenient, but not to be ignored, the wish to go to Antarctica was suddenly there. "Still, the thought was there. Antarctica… I have not always longed to go to Antarctica, or even ever wanted to especially, but the thought was as powerful as if it had been a lifelong dream. Perhaps it's possible to have lifelong dreams in retrospect". She needed a place, which only could exist in her mind, a place where there would be no thought, no pain, nor stimulating colours; a place, which reminded her on previous stays in psychiatric facilities; a place where she could find the same kind of solace, but without the annoying presence of the nurses: "I wanted white and ice for as far as the eye could see, and I wanted it in the one place in the world that was uninhabited (never mind the penguins, seals and base camp personnel for the time being). I wanted a place where Sister Winniki couldn't exist. I wanted my white bedroom extended beyond reason. That was Antarctica, and only Antarctica."

Keneally (2001) understands his obsession with the open white spaces as a metaphysical experience: "This is the icy Eden many modern readers consider their favourite mental landscape on earth…the South Pole, a place where all is north, where the world can be circled in four steps, a point as absolute as some mystic's conception of the Deity." Soul-searching and Adventure Tourism sometimes go together. There is a prolific production of extreme sport events, privately organized expeditions and related to both of these the publication of their accounts. Their adventures capture the will and fear of Antarctic exploration. And however diverse there expeditions might be, they all do establish a dialogue with the heroes of their childhood.

The most famous memoir of the 1910–1913 British Antarctic Expedition led by Robert Falcon Scott was written and published in 1922 by one of its members, Apsley Cherry-Garrard. The book earned wide praise for its frank treatment of the difficulties of the expedition, the causes of its disastrous outcome, and the meaning of human suffering under extreme conditions. Cherry-Garrard's masterpiece, "The Worst Journey in the World", refers in the first place to a winter journey he took together with two other members of the expedition from Cape Evans to Cape Crozier. Their goal was to recover eggs of the Emperor penguin for scientific study. It was thought at the time that the flightless penguin might shed light on an evolutionary link between reptiles and birds through its embryo. As the birds nest during the Antarctic winter, it was necessary to mount a special expedition in July 1911. Complete darkness and temperatures of −40 °C and below made this "the worst Journey in the World". But they succeeded and Cherry-Garrard later accompanied the initial team across the Ross Ice Shelf and up the Beardmore Glacier. At the edge of the polar plateau, Scott sent him and other men, who were not chosen to go on to the Pole to return to the base camp at Cape Evans. Cherry-Garrard had been given the task of using the dog teams to meet Scott's party and assist them home, but he did not penetrate beyond One Ton Depot, only 11 miles distant from Scott's final location where he and his companions froze to death. "*The Worst Journey in the World*" asks, but does not answer, the question of whether this suffering was futile, or whether it would inspire future human beings facing very different challenges. Modern day debates unfold around these questions and Cherry-Garrard is still a much frequented source for polar travellers. In 1994, his book was published as the first numerical entry in the Picador Travel Classics. The July/August 2001 issue of National Geographic Adventure listed the "The 100 Best Adventure Books of All Time", with "The Worst Journey in the World" named first. A drama documentary, under the same title, was broadcast on BBC Four in April 2007, which was followed by a two-part radio drama in the Classic Serial strand, first broadcast in September 2008. With typical British understatement Cherry-Garrard would exclaim that "Polar exploration is at once the cleanest and most isolated way of having a bad time" (Cherry-Garrard 1922).

Even today there are still places in which no human has set foot, mountains have yet to be conquered, ways of crossing Antarctica, which have yet to be devised. There are still new records to be achieved and although this continent is not for the faint of heart, as literature has well established, the challenges are out there. For mountaineers a must in their reading list, Messner's, "Antarctica: both Heaven and Hell" (1992), deals both with the physical challenges and the more spiritual questions involved in polar expeditions. He recounts the first crossing on foot, together with Arved Fuchs, of Antarctica in 92 days ending in mid-February, 1990. His treatment of the white void in front of him as a meditative medium becomes the centre of his reflections about how to endure in this harsh environment.

Felicity Aston, physicist and meteorologist, took 2 months off from all human contact as she became the first woman—and only the third person in history—to ski across the entire continent of Antarctica alone. Aston's journey across the ice at the bottom of the world asked of her the extremes in terms of mental and physical bravery, as she faced the risks of unseen cracks buried in the snow so large they might engulf her and hypothermia due to brutalizing weather. She had to deal, too, with her emotional vulnerability in face of the constant bombardment of hallucinations brought on by the vast sea of whiteness, the lack of stimulation to her senses as she faced what is tantamount to a form of solitary confinement. Her account of a spectacular and appalling journey, "Alone in Antarctica: The First Woman to Ski

Solo across the Southern Ice" (2014) was applauded by the media and praised for its power to inspire others to follow their own dreams. She took great risks, just like the polar explorers of the past had done. "Generally the risks were taken, for, on the whole, it is better to be a little over-bold than a little over-cautious, while always there was a something inside urging you to do it just because there was a certain risk, and you hardly liked not to do it. It is so easy to be afraid of being afraid!" (Cherry-Garrard 1922).

Another modern day adventurer and writer, much acclaimed by the press and other media was Alastair Vere Nicoll with his book "Riding the Ice Wind: By Kite and Sledge across Antarctica" (2010). Leaving the security of friends, work and a wife, Vere Nicoll joined a team of young men to harness the katabatic winds and haul and kite-surf across Antarctica. His story is not only about the first West-to-East traverse of the continent of Antarctica, but of the crossing of two phases in his life—from youth into manhood, fantasy into reality. It is also the story of a race against time, as he fought to get home for the birth of his first child. As Alastair battled through the freezing wastes, exploring the earth's wildest continent and his deepest self, he was haunted by ghosts of past explorers and by the question of what it is to be a "modern man". John Hare, author of "Mysteries of the Gobi", would later say that "It's extremely heartening, to discover, that a younger generation of adventurers has got what it takes—and more. They prove themselves worthy successors to their heroes, Amundsen, Shackleton and Scott" (2009).

For others, the waters surrounding Antarctica are more appealing than the ice ashore. In January 2001 three men from New Zealand set out to paddle down the length of the Antarctic Peninsula. It was to be the southernmost sea kayak journey ever attempted, a 528-mile expedition through the freezing waste of ice, rock, and ocean that makes this one of the most inhospitable coasts on earth. Their book, "The Frozen Coast: Sea Kayaking the Antarctic Peninsula" (Charles 2004), offers a gripping account of this ultimately successful journey and at the same time invites future adventurers to follow in their footprints, by offering a broad section about the equipment needed. And once again it testifies to the physical challenge and the hardships the kayakers had to overcome by careful preparation, ingenuity and determination. And while the three intrepid kayakers tried to stay above the water, others decided, that it was time to explore the underwater world of Antarctica.

Until recently the seventh continent has been the exclusive realm of scientific and military divers. Today, however, the icy waters of Antarctica have become the extreme destination for recreational divers wishing to explore beyond the conventional. They hope to observe the strange marine life that abounds below the surface, feel the thrill of meeting Leopard seals one to one and dive under the ice. Kelley Eareckson's "The Antarctic Dive Guide" (2015) is still the first book offering information about both the history of diving in Antarctica and its wildlife. Essential information is also provided on how to choose and prepare for travel to this remote region, and diving techniques for subzero waters. Antarctica's underwater world is little-explored, which makes it even more attractive for visitors. More and more tourists wish to dive in Antarctica. And by doing so, they live up to their own expectations reaching physical and psychological limits outlined in adventure tourism excursions (IAATO 2014a).

Extreme athletes take this challenge to a whole new level. Their life stories inspire readers equally with their drive and their fierce determination. The British swimmer Lewis Pugh broke two world records, for the most southerly swim ever undertaken in the ocean, and the longest-duration polar swim ever completed. On 14 December 2005, Pugh swam a kilometre in the seas off the Antarctic Peninsula at a latitude of 65° south, some of the world's coldest waters, where the sea's saltiness allows temperatures to dip to just below 0 °C without freezing. The feat, which took 18 min and 10 s, required him not only to maintain a safe body temperature throughout the ordeal, but also to stave off the crippling effects of the body's natural reaction to icy water. Two days later, he swam a mile off the nearby Deception Island, spending 30 min and 30 s in the water—longer than any other polar swimmer. How can a person wearing only Speedo trunks survive the icy sea? Less than a second after hitting the water, the lungs constrict, causing an untrained person to hyperventilate, taking up to 60 gasping breaths per minute. Heart rate and blood pressure also skyrocket as the heart panics in response to the shock. Pugh says he felt "screaming pain" all over his body as soon as he dived in. The paralysing effect of this response means that, in icy water, the body becomes its own worst enemy and the mind has to make up for it with sheer determination. Mental imagery can have a profound effect on physical responses, and Pugh seems to have trained for this occasion such a long time, that he is capable of generating more body heat, when needed. The cold-water swimmer Lynne Cox got her first taste of extreme swimming in a pool, when she was only 9 years old: "My world was reduced to the blur of my arms stroking as a cold, driving rain began. The raindrops that hit my lips tasted sweet and cold, and I enjoyed the sensations of every new moment. The pool was no longer a flat, boring rectangle of blue; it was now a place of constant change, a place that I had to continually adjust to as I swam or I'd get big gulps of water instead of air. That day, I realized that nature was strong, beautiful, dramatic, and wonderful, and being out in the water during that storm made me feel somehow a part of it, somehow connected to it … I realized that by putting myself in a situation different from everyone else's, I had experienced something different, beautiful, and amazing" (Cox 2004). After this first glimpse of stardom it is not surprising, that she went on training for more adrenaline. The American long-distance open-water swimmer and writer became the first person to swim the Straits of Magellan, the first to swim around the Cape of Good Hope. And another of her accomplishments was swimming more than a mile (1.6 km) in the waters of Antarctica. Cox was in the water for 25 min, swimming 1.22 miles (1.96 km). Her book about the experience, "Swimming to Antarctica", was published in 2004. The health benefits of these swims are doubtful, the risks less so. And yet every year, some people swim in polar waters. It seems to be almost like a spiritual experience to some and Cox's book gives insight to the mental exercise it involves. One thing that encourages people to endure the pain is the accompanying cocktail of endorphins that arises in the brain, resulting in a lasting sense of euphoria and calm. "It sets you up for the day," But a more mental problem seems to be the fear of losing one's life. A problem historic polar heroes knew all too well. "And if the worst, or best, happens, and Death comes for you in the snow, he comes disguised as Sleep, and you greet him rather as a welcome friend than a gruesome foe" (Cherry-Garrard 1922). Cox tries a different approach by offering advice on how to deal with the fear and ultimately guarantee the survival of the swimmer this way. "I'm writing about the Fear because it's a common problem. My theory is that you can't get rid of it, but you can have a few tricks to deal with what feels like a life-threatening situation, particularly if there isn't another more-frightened swimmer to accompany. So far in my repertoire I have the following, but I'd welcome new ideas: breaststroke a while, concentrating on getting your breath under control, bringing your heart rate back to normal. Go on to your back and float a little, if you can. Look at what's above and around you."

Antarctica with its extreme geography certainly offers possibilities to look around. But there is more reasons to choose the Seventh Continent for the special experiences, travellers are looking for. Antarctica is a space that we tend to configure according to our own deepest preoccupations whether we have been there or not. And then there are of course images that break through our unconsciousness, quite uninvited, as if they have been there for a long time, lurking. The skipper and writer Alvarez Forn (1991), who sailed on board the sailing vessel "Pequod" to Antarctica might stand as an example for others: "It was in April 1987, when a subconscious idea broke to the surface, which must have formed quite a while ago, when we came back from our voyage to Cape Horn. It was suddenly there, I had to go to Antarctica, without much further ado or measuring the consequences, I just knew that I had to sail south. I wasn't competing with anybody, but it seemed to be an additional bonus, that the first sailing yachts, which had wanted to go in the years 82 and 83, had not succeeded."

Many others feel the same urge to excel and compete in expeditions. The Irish Pat Falvey is one of them. Having completed the Seven Summits Challenge twice, his adventures have taken him to some of the highest, coldest, loneliest and most remote places on earth. He reveals the danger and joys at the limits of physical and mental challenges on an Antarctic Expedition in his book "*A Journey to Adventure. Stories I never thought I'd tell*" (2007). The traverse of South Georgia and a visit to Elephant Island were part of his voyage to Antarctica. "The highlight of our expedition was the moment we reached Cape Valentine, where Shackleton first landed to rescue his crew. We paid homage to our heroes by landing on the very spit of shore where the crew awaited their boss so many years ago. For me it was a dream come true to stand in a place that had been etched in my memory from photos I'd seen and books I had read. I was so proud of our team for coming to Antarctica to pay homage to Irishmen who had achieved an incredible feat of survival. As I left Elephant Island, my journey was just beginning. Next, I would set out to achieve Shackleton's dream of reaching the South Pole and beyond. But that will be another story."

References

Alvarez Forn, H. (1991). *Antarktikos* (310p.). Buenos Aires: Biblioteca de a bordo.
Arthur, E. (1994). *Antarctic navigation* (700p.). New York: Alfred Knopf.
Aston, F. (2014). *Alone in Antarctica: The first woman to ski solo across the Southern Ice* (320p.). Counterpoint.
Bainbridge, B. (1993). *The birthday boys* (189p.). London: Penguin Books.
Batchelor, J. C. (1981). *The birth of the people's republic of Antarctica* (401p.). New York: Dial Press.
Botaya, F. (2010). *Antártida 1947* (382p.). Madrid: Nowtilus.
Captain Seaborn, J. (2009). *Symmzonia: A voyage of discovery* (118p.). Kindle Edition, (first pub. 1820).
Charles, G. (2004). *The Frozen Coast: Sea Kayaking the Antarctic Peninsula* (120p.). The ions Press.
Cherry-Garrard, A. (1922). *The Worst Journey in the World*. London.
Cox, L. (2004). *Swimming to Antarctica* (384p.). Alfred A. Knopf.
Diski, J. (1997). *Skating to Antarctica* (250p.). London: Granta Books.
Eareckson Kelley, L. (2015). *The Antarctic Dive Guide* (144p.). Princeton University Press.
Falvey, P. (2007). *A Journey to Adventure. Stories I never thought I'd tell* (170p.). Collins Press.
IAATO. (2014). *ATCM XXXVII-CEPCVII, Information Paper No 044, Overview of Antarctic Tourism.* www.ats.aq
IAATO. (2014a). *ATCM XXXVII-CEPXVII, Information Paper No 078 Adventure Tourism: Activities undertaken by IAATO Members Brazil 2014.* www.ats.aq
IAATO. (2015). *ATCM XXXVIII-CEP XVIII, Information Paper No 053, Overview of Antarctic Tourism.* www.ats.aq
Jarvis, T. (2014). *Chasing Shackleton. Recreating the world's greatest journey of survival* (272p.). William Morrow.
Keneally, T. (2001). *Victims of the Aurora* (219p.). New York: Harcourt.
Kitto, C. (1983). *The Antarctica cookbook* (187p.). London: Duckworth.
Leane, E. (2012). *Antarctica in fiction. Imaginative narratives of the far south* (235p.). New York: Cambridge University Press.
Leane, E., Crane, R. & Williams, M. (2011). *Imagining Antarctica. Cultural perspectives on the southern continent* (121p.). Tasmania: Quintus Publishing.
Lovecraft, H. P. (2005). *At the mountains of madness.* New York: Modern Library, (first pub. 1931).
McCaughrean, G. (2011). *The White Darkness* (262p.). New York: Oxford University Press.
Melville, H. (1992). *Moby Dick.* New York: Penguin, (first pub. 1851).
Messner, R. (1992). *Antarctica: Both Heaven and Hell* (381p.). Mountaineers Books
Mickleburgh, E. (1988). *Beyond the frozen sea: Visions of Antarctica* (256p.). St. Martins Press.
Morrell, B. (2014). *Narratives of four voyages to the South Sea, ... and Antarctic ocean* (506p.). Cambridge University Press, (first pub. 1832).
Morris, M.E. (1988). *The icemen. A novel of Antarctica* (330p.). Novato CA: Presidio.
Poe, E. A. *Manuscript found in a bottle*, (first publ. 1833).
Poe, E. A. (1999). *The narrative of Arthur Gordon Pym of Nantucket.* New York: Penguin, (first pub. 1838).
Reynolds, J. N. (1836) *Address on the Subject of a Surveying and Exploring Expedition to the Pacific Ocean and the South Seas.*
Robinson, K. S. (1997). *Antarctica* (685p.). London: Harper Collins.
Robinson, J. (2009). *La resurrección de Antártica* (446p.). Barcelona: ViaMagna Ediciones.
Spufford, F. (1997). *I May be Some Time. Ice and the English imagination* (357p.). New York: Picador.
Taylor Coleridge, S. (1970). *The Rime of the ancient mariner* (77p.). Dover Publications, (first pub. 1798).

Trojanov, I. (2011). *Eistau* (176p.). Frankfurt: Deutscher Taschenbuch Verlag.
Vere Nicoll, A. (2010). *Riding the Ice Wind: By Kite and Sledge across Antarctica* (272p.). Tauris.
Verne, J. (2005). *Voyage to the centre of the earth* (160p.). Dover Thrift edition, (first pub. in French 1864).
Verne, J. (2005). *An Antarctic Mystery, or, the Sphinx of the Ice Fields.* Rockville, MD: Wildside, (first pub. in French 1897).
Wainschenker, P. (2013). *La imaginación y la Antártida. Representaciones de las regiones polares del sur en cine y literatura. Tesina de grado, Ciencias de la Comunicación.* Universidad de Buenos Aires (mimeo).
Weddell, J. (2011). *A voyage towards the South Pole: performed in the years 1822–24* (316p.). Cambridge University Press, (first pub. 1825).
Wheeler, S. (1996). *Terra incognita. Travels in Antarctica* (302p.). London: Vintage.

Chapter 3
Supervision of Antarctic Shipborne Tourism: A Pending Issue?

Rodolfo A. Sánchez and Ricardo Roura

Abstract This chapter focuses on the ways Parties to the Antarctic Treaty, and the tourism industry supervise the implementation of applicable regulation for Antarctic shipborne tourism. Tourism regulation is adopted in the framework of the Antarctic Treaty Consultative Meetings, as well as the external provisions from other international fora and by industry self-regulation. Based on a description of the existing regulation and provisions in place, this chapter discusses the way supervision of Antarctic tourist operations have been debated, conducted and reported by both Treaty Parties and the industry. The article concludes by discussing likely mechanisms to enhance supervision of the management of shipborne tourist activities in Antarctica, and proposing further lines of research on this issue.

Keywords Antarctic tourism · Antarctic treaty system · Tourism regulation · Tourism supervision · Visitor guidelines · Port state control · Flag state control

3.1 Introduction

In conjunction with the rapid growth and diversification of Antarctic tourism since the late 1980s, and every decade since then, much emphasis has been put by States Parties to the Antarctic Treaty and academics on the need for regulating tourism activities and monitoring their effects on the Antarctic environment. Environmental non-governmental organizations have also addressed this issue critically. However,

R.A. Sánchez (✉)
Dirección General de Asuntos Antárticos, Ministerio de Relaciones Exteriores y Culto, República Argentina, Esmeralda 1212, C1007ABR Ciudad Autónoma de Buenos Aires, Argentina
e-mail: zrf@mrecic.gov.ar

R. Roura
Antarctic and Southern Ocean Coalition, 1320 19th street, NW, 5th floor, Washington D.C. 20036, USA
e-mail: ricardo.roura@worldonline.nl

M. Schillat et al., *Tourism in Antarctica*,
SpringerBriefs in Geography, DOI 10.1007/978-3-319-39914-0_3

issues connected to supervision of Antarctic tourism activities have been much less addressed by any of these groups.

This chapter addresses aspects of Antarctic tourism supervision with a focus on shipborne tourism, which is the dominant form of conducting tourist activities in Antarctica. Some of these aspects have been addressed in the existing available literature; however, such studies were mainly focused on the content of existing regulation, rather than on its actual practical implementation. This study contributes therefore with a new approach, based on the methods used by Antarctic Treaty states and the tourism industry to enforce the existing rules and regulations.

The concept of supervision may comprise a number of different activities and actions. In this paper, supervision refers to all activities and actions carried out by the Antarctic Treaty Consultative Parties (ATCPs) and the Antarctic tourism industry that aim to regulate the activities of tour operators, individually and collectively.

For ATCPs, supervision includes desktop authorization/assessment processes; and inspections/observer programs aimed to assess the level of compliance of existing regulation, either at port facilities, on-board of cruise ships en route to Antarctica or at sites where tourism landings take place. The legal basis for each of these means varies from country to country and generally for ATCMs. Supervision by the Antarctic tourism industry also includes desktop assessment processes and observer programs, though their focus and extent may, on occasions, be different to those of the ATCPs. The legal basis for industry regulation also differs from those used by governments. Overall, this chapter will focus primarily on inspection/observer programs, although there will also be some references to desktop processes as well.

Capacity building activities, which include all actions aimed to enhance knowledge, skills and capabilities of tour operators and their staff, in order to ensure their awareness of existing regulations and provisions, lie outside the focus of this work. Environmental monitoring aimed to assess the cumulative effects of tourism operations at selected locations will not be considered here either.

This analysis is based on our experience as stakeholders in the tourism debate as representatives of a State Party (Sánchez 1999–2011) and environmental organizations accredited as experts to Antarctic Treaty fora (Roura 1997-present). Our experience is complemented with documentation emerging primarily from Antarctic Treaty fora, which reflects the "frontline" of the tourism discussion, rather than on academic analysis of this issue.

3.2 Tourism in the Antarctic Context

Traditionally, Antarctic tourism has been characterized by ship expedition cruises departing from the southern part of South America to the Antarctic Peninsula from about October to March, with activities that include landings and operation of small boats with the primary purpose of sightseeing. Shipborne tourism operations also take place in the Ross Sea. More recently, these include limited visitation to East Antarctica destinations as well (Jabour 2014).

A rapid development of tourism in Antarctica has occurred since the late 1980s, with increasing visitor numbers and a diversifying supply of transport modes and activities (Lamers et al. 2008). While some 5000 tourists visited Antarctica in the 1993/1994 season, around 37,000 passengers did it so during the austral 2013–2014 and 2014–2015 seasons (IAATO 2015). Diversification of Antarctic tourism occurred by the incremental introduction of other types of tourism operations like cruise-only itineraries without landings for ships carrying more than 500 passengers; fly–sail operations; yacht cruising; over flights; and adventure tourism activities such as kayaking, scuba diving, mountain climbing snorkeling, skiing, snowboarding, camping (including short overnight stays), running events, helicopter operations and underwater ROVs (Lamers et al. 2008). Versions of some of these activities are now conducted as part of standard shipborne tourism as a complement of sightseeing.

These changes have in turn resulted *inter alia* from inter-industry competition, as well as expansion and segmentation of the tourism market.

ATCPs have expressed concerns about tourism developments since early on, with ATCM Recommendation ATCM IV-27 (Santiago 1966) *Regulation of Antarctic Tourism* addressing concerns about disruption at research stations by tourism visits. More recently, many authors, as well as Antarctic Treaty Parties, have expressed concern over the growth of Antarctic tourism in particular since the late 1980s (see e.g. Hall 1992; Cessford and Dingwall 1996; Bastmeijer and Roura 2004; The Netherlands 2014), mainly due to the potential environmental implications of a wider range of tourism activities, and of a geographical tourism expansion. Considerable attention has also been paid to the increasing potential for serious incidents and accidents, as ships operating in the Antarctic environments are exposed to Antarctic-specific risks, including extreme weather conditions, incomplete/inadequate charting and limited coverage of communication systems and other navigational aids[1]

3.3 The Stakeholders of Antarctic Tourism

The Antarctic Treaty was signed in 1959 and entered into force in 1961. At present, membership consists of 29 Consultative Parties (ATCPs or Antarctic Treaty Parties), which are entitled to participate in the process of consensus decision-making. In addition, there are 24 Non-Consultative Parties, which do not participate in decision-making. The Antarctic Treaty Parties conduct their formal

[1]Concerns on ship safety increased especially after the various reported incidents and accidents that happened between 2007 and 2009, including the well-publicized sinking of the M/S Explorer in Bransfield Strait off the Antarctic Peninsula in 2007 (ATS 2014; Liggett et al. 2011).

business at annual meetings known as Antarctic Treaty Consultative Meetings (ATCMs), which are held for the purpose of "exchanging information, consulting together on matters of common interest pertaining to Antarctica, and formulating, considering, and recommending to their Governments, measures in furtherance of the principles and objectives of the Treaty"(Antarctic Treaty, Article IX).

The ATCMs normally consist of several consecutive or parallel sessions: the Plenary; two or more Working Groups (WGs); and since 1998 also the meeting of the Committee for Environmental Protection (CEP). Currently, two Working Groups function in the ATCM's framework: WG1 deals with policy, legal and institutional issues and WG2 have responsibility for operations, science and tourism. It should be noted that between 2004 and 2015 there was a dedicated Tourism and Non-Governmental Activities in which all issues connected to Antarctic tourism—including proposals for new regulations—were discussed. This Working Group was terminated in 2015 and these discussions will now take place on a new Working Group 2.

Aside Treaty Parties, a number of Observers and Experts are invited to participate in the ATCMs. One of the expert organizations most relevant for this study is the International Association of Antarctica Tour Operators (IAATO), a coalition of tour industry affiliates founded in 1991 to advocate, promote and practice safe and environmentally responsible private-sector travel to the Antarctic. IAATO's founding objectives include operating within the parameters of the Antarctic Treaty System, as well as the other applicable international and national instruments, notably regulation by the International Maritime Organization (IMO). As a whole, most of the Antarctic shipborne tour operators are members of IAATO: its membership currently consists of 45 companies who operate travel programs to the Antarctic and/or sub-Antarctic islands that have fulfilled the IAATO membership requirements set up by its internal by-laws. By 2016, the IAATO vessel fleet included 34 ships (13 to 500 + passengers) and 21 yachts (up to 12 passengers) (www.iaaato.org, accessed March 12th 2016).

3.4 Current Tourism Regulation

The two main objectives for regulating Antarctic tour operations are to ensure safety of operations, and to minimize the environmental impacts arising from these activities. Primarily, regulation aims to avoid marine pollution from ships, and to ensure minimal disturbance to Antarctic flora, fauna and cultural heritage, as well as scientific research.

The current regulatory framework related to Antarctic tourist activities consists in the first place on the provisions of the Antarctic Treaty System (ATS), which include the Antarctic Treaty, the Protocol, and legally binding measures, procedural decisions and hortatory resolutions agreed at ATCMs. The first ATS regulation on tourism dates back to 1966, when Recommendation IV-27 "Effects of Antarctic Tourism" was agreed.[2] By 1975, Treaty Parties already acknowledged, through Recommendation VIII-9 "Statement of Accepted Practices in Tourism", that tourism required regulation. Since then and up to 1994, when a very general set of "Guidelines for tourism" (for visitors and tour operators) were approved through Recommendation XVIII-1, very few recommendations on tourism had been agreed to at ATCMs.

The Protocol on Environmental Protection to the Antarctic Treaty (hereafter the Protocol), agreed in 1991 and in force since 1998, applies to all activities south of 60° South. The Protocol is the main framework for environmental protection in Antarctica and is applicable to tourism although it does not regulate tourism specifically. After the Protocol, activities in the Antarctic are subject to a number of obligations and procedures related to Environmental Impact Assessment, protection of flora and fauna, waste management, marine pollution from ships and protected areas.

The ATS regulation on tourism that came after the Protocol consists, generally speaking, of codes of conduct or behavioural guidelines (mainly the non-legally binding Site Guidelines for Visitors) and reporting requirements connected to the exchange of information system in place within the ATS. Other issues covered by ATS regulatory provisions applicable to tourism operations also include guidance and tools to assist in the conduct of different aspects of tourist activities.

Because the vast majority of Antarctic tour activities are ship-based, regulation comes also from the corpus of existing international maritime law via flag states (Jabour 2011). This includes the International Convention for the Prevention of Pollution from Ships (hereafter MARPOL 1973/78), the International Agreement Concerning Safety of Life at Sea (hereafter SOLAS 1974), and the International Convention on Standards of Training, Certification and Watchkeeping for Seafarers (STCW 1978/1995), among others (http://www.imo.org).

The recent adoption (in 2011 and 2014) of new IMO regulations is to dramatically affect the way Antarctic cruise ship operations are carried out. An amendment of MARPOL Annex I banning the use or carriage of heavy fuel oil in the Antarctic entered into force in 2011, limited the operation of large cruise-only vessels in Antarctic waters. In addition, the International Code for Ships Operating in Polar Waters (Polar Code) was adopted in November 2014, together with related amendments to the International Convention for the Safety of Life at Sea (SOLAS) to make it mandatory. The Polar Code is intended to cover the full range of shipping-related matters relevant to navigation in polar waters including ship design, construction and equipment, operational and training concerns, search and rescue, and the protection of the polar environments. The Code would then require ships intending to operate in Antarctic (and Arctic) waters to apply for a Polar Ship Certificate, which will be issued once a thorough assessment of the vessel is carried

[2]This recommendation set up a permitting system for groups of tourists visiting Antarctic stations.

out.[3] The adoption of the Polar Code marks a historic milestone in the IMO's work to protect ships and their passengers in polar waters, and will certainly have an impact in the safety of tourism vessels in Antarctic waters and in the conduct of their operations as well (http://www.imo.org).

This bulk of legislation is complemented by industry self-regulation through IAATO. Over time IAATO has prepared a number of guidelines and procedures that members should observe, in conjunction to its internal by-laws, some of which have constituted the basis for the ATS own regulations[4] (Jabour 2014).

It has been argued that the formal ATS regime was not sufficient, as it seemed to lag behind the level of growth and diversification that Antarctic tourism had experienced (Scott 2001; Bastmeijer and Roura 2004; Molenaar 2005; Amelung and Lamers 2006; Lamers et al. 2008). This criticism is based on the fact that regulation about tourism stemming from the ATS consists mainly of non-legally binding Resolutions; and two legally binding Measures,[5] neither of which has yet entered into force. However, it has been argued that the ATCPs' engagement with IMO on the preparation of the Polar Code—which has been finally adopted—will probably help counter much existing criticism of the effectiveness of tourism regulation (Jabour 2014).

Despite this, some regulatory gaps may still remain and Treaty Parties seem to have recognized the need for further legislation. When in 2009 Resolution 7 (General Principles of Antarctic Tourism) was agreed, the ATCM recommended ATCPs "to ensure, as far as practicable, that they continue to proactively develop regulations relating to tourism activities that should provide for a consistent framework for the management of tourism".

3.5 Who Is Responsible for Supervising?

Irrespective of the opinions about as to whether Antarctic tourism regulation is currently sufficient (or will still be sufficient in the near future), there seems to be already quite a large corpus of mandatory and non-mandatory provisions. In theory, if properly fulfilled, these provisions would ensure that Antarctic tourism was carried out safely and in a moderately environmentally responsible manner. However, as noted above, tourism is a dynamic industry and new modalities have

---

[3]The expected date of entry into force of the Polar Code is 1 January 2017, and it will apply to new ships constructed after that date. Ships constructed before that date will be required to meet the relevant requirements of the Polar Code by the first intermediate or renewal survey, whichever occurs first, after 1 January 2018 (http://www.imo.org).

[4]For instance, Measure 15 (2009) which establishes certain restrictions for the landing of persons from passenger vessels was based on a previous IAATO provision.

[5]These were Measure 4, 2004, "Insurance and Contingency Planning for Tourism and NGO activities in the AT Area" and Measure 15, 2009, "Landing of Persons from passenger Vessels in the AT Area".

developed ahead of regulation, sometimes very rapidly.[6] In addition, enforcement of regulations is still a challenge. The uniqueness of Antarctica as a tourist destination causes regulatory problems, as there is no single sovereign entity than can make and enforce a uniform set of laws relating to tourist activities in, or affecting its territory in Antarctica that can be universally applied to every tourism vessel, operator or tourist (Jabour 2014). So, who should be responsible for ensuring such a corpus of regulation is adequately observed?

Here, we should distinguish a first layer of generic regulation relevant to Antarctic shipborne tourism that derives from the Antarctic Treaty System, such as the Protocol, from that which derives from other relevant international instruments. A second layer is regulation specific to tourism. Any of these forms of regulation can be binding or non-binding. In parallel there is industry (i.e. IAATO) self-regulation, which is non-binding and only applicable to IAATO members. Although most tour operators responsible for cruise ships in Antarctica are currently affiliated to IAATO, there is no warrantee that this will always be the case (Liggett et al. 2011; Lamers et al. 2012).

In this context, it is assumed here that in general ATCPs should supervise the tourism—that is, primarily the collective of tour operators; and that tour operators should supervise tourists.[7] How does this work for shipborne tourism?

First, ATCPs are responsible to ensure tour operators registered in their territories comply with the Protocol's Environmental Impact Assessment requirements and with those norms agreed by ATCMs. The latter includes not only checking that exchange of information requirements are observed (through appropriate national assessment/authorization desktop procedures) but also ensuring that supervision of on-board and in-field tourist operations is regularly carried out.

A number of general and particular norms stemming from ATCMs regulate the information on tourist operations that ATCPs should exchange. Treaty Parties in turn should obtain such information from tour operators registered in their countries. According to Resolution 6 (2001) *Information Exchange to be carried out through central website according to Information Exchange Requirements*, ATCPs should submit every year information on planned tour operations (pre-season information), and an update of such after the end of the austral summer season (annual report). This encompasses information on vessel-based operations, including data on the operator and the ship, as well as on their Antarctic itineraries.[8]

---

[6]For instance, the issue of Unmanned Aerial Vehicles (UAVs) had never been discussed at the ATCM until 2014, when it required attention as it was creating some problems, including for tourism. Subsequently IAATO developed regulations for this activity, preventing tourists in coastal areas to use this technology (but not inland, and not for some commercial activities under permit).

[7]Inter-company supervision is here excluded. In addition, IAATO can supervise individual members according to its own observer mechanisms.

[8]After Decision 4 (2012), it was also decided to give Parties, as applicable, the option to include the denial of authorisations to operators and identify activities cancelled by an operator after meeting Parties' regulatory requirements.

As a general principle, the ATCM recommends Treaty Parties to make that information publicly available.

In addition, according to Resolution 1 (2005), ATCPs should also annually provide Antarctic Treaty Secretariat with a list of the Initial Environmental Evaluations (IEEs) and Comprehensive Environmental Evaluations (CEEs) prepared by or submitted to them. This list should include a brief summary of their relevant data in order to be posted on the Secretariat's website (www.ats.aq). A copy in electronic format of these documents should also be submitted, where possible.

However, the way pre-season and annual cruise ship information actually correlate is still unknown, and the level of disparity of the information presented by different ATCPs under the mechanisms adopted by Resolution 1 (2005) still seem to be significant. For instance, while some Parties report and make the prepared IEEs available, others only report them and have submitted very few, if any, IEEs as part of the Exchange of Information System

In addition, it is up to the operators to ensure that activities carried out on the ground do follow what the requirements as written in the EIA. In other words, the theory of EIA may not correspond to the reality on the ground (Hemmings and Roura 2003; Bastmeijer and Roura 2008).

In brief, do ATCPs know what tour operators do in Antarctica? Actually, they do (or should) know what tour operators under their supervision (i.e. those who have a legal address registered in their territories, or flying their flags) do. As far as the remaining tour operators are concerned, ATCPs know more now than they used to know in the past, as more EIAs are becoming available through the Information Exchange System. When this is not the case, ATCPs only know that such tour operators' activities have at least gone through a National authorization/assessment process.

In this sense, concern has been expressed at ATCMs about the different legal systems of different ATCPs, which may provide a variance in the interpretation of Protocol requirements and their implementation. According to some ATCPs, this could lead to proponents (especially those involved in adventure tourism activities) "forum shopping" in various jurisdictions in order to find the easiest way to proceed (ATS 2013: 231). Though providing copies of IEEs is not specifically an obligation for Treaty Parties, it would help to identify—and to a degree prevent—inconsistencies among different legal regimes. It would also contribute to build a basis for supervision of tourism and increase the level of transparency of the exchange of information system in place.

It has also been argued that, in relation to tourist activities, compliance-oriented transparency (closely linked to compliance monitoring) was inadequate and that improvements within the ATS would be desirable. It also has been proposed that while there would be a continuous need for adequate transparency within the Antarctic tourism regulatory regime, it seemed to be a matter of political will, dedication and incentives to improve the current reporting, information-sharing and communication networks within and between both the ATS and IAATO (Haase 2008).

Second, ATCPs and other States should commit to assess whether ships flying their flags undertake their shipping operations in accordance with IMO regulations.[9] FlagState Control is the authority an administration has over vessels under its registry (flag) regardless of where they are operating. The United Nations Convention on the Law of the Sea (UNCLOS) regulates the duties of the flag State in High Seas through its article 94. In a nutshell, every Flag State has to "effectively exercise its jurisdiction and control in administrative, technical and social matters over ships flying its flag".

In third place, there is Port State control, which can be described as the inspection of foreign registry ships in national ports by port authorities in departure states, for the purpose of verifying that the condition of the ship and its equipment comply with the requirements of certain international maritime conventions and that the ship is manned and operated in compliance with applicable national laws (New Zealand 2009a, b). The UNCLOS stipulates the attributions of the port States to enforce legislation on the protection and preservation of the marine environment through its article 218.[10]

Port State Control was originally intended to be a back up to flag State implementation. Despite the fact that the primary responsibility for ships' standards rests with the flag State, Port State Control provides a "safety net" to catch substandard ships (www.imo.org). It has been argued that this was due to doubts about the effectiveness of some flag States, given their inability to establish a competent administration (Plaza 1999; Valenzuela 1999). In addition, a number of national flags prominently appear in the upper reaches of the lists of port State control detentions, insurance loss and casualty figures and the like (Plaza 1999). Alternative approaches that have the possibility of ensuring a more effective implementation of rules and regimes in terms of port state jurisdiction have therefore emerged (Orrego Vicuña 2000), and IMO has recognized the potential for such approaches, especially if Port State Control is organized on a regional basis, through a number of existing Memoranda of Understanding which already cover all of the world's oceans (www.imo.org).

Finally, IAATO is to control that its Members[11] duly comply with its by-laws and internal procedures in place. In this respect, for IAATO operators membership this encompasses, as a first step, a desktop review of the operator's activities, including a preliminary review by the IAATO Membership Committee (IAATO 2011).

---

[9]All Antarctic Treaty Parties, with the exception of Belarus, are also IMO Parties.

[10]In addition to pollution prevention requirements the PSC inspection will focus upon safety issues by verification that the vessel's crew meet international standards, that it has a functional safety management system, that its charts are corrected and up-to-date, that it has undertaken passage planning for the duration of its time in Antarctic waters, that it has adequate lifesaving equipment and the means of making distress signals and that it has a search and rescue plan in place and lodged with a rescue co-ordination centre (New Zealand 2009a, b).

[11]The operation of expedition cruise ships exclusively takes place within the IAATO-framework, while during the time period from 1996 to 2013 merely 16 % of a total of over 200 yachts were IAATO-members during the time of their Antarctic voyage (Vöneky 2016).

For cases of non-compliance of its regulation, as of 2013 IAATO has implemented Rules of Procedure for compliance and dispute resolution. IAATO has set up a mechanism to consider and verify alleged violations to its by-laws and regulations (which can be reported/documented either by operators themselves or by a third party). Such a mechanism is activated if the incidents are repeatedly committed and believed to be deliberate. Potential sanctions to non-compliant members include, among others, removing them from IAATO website, and or reporting them to their competent authority/flag state. IAATO has no capacity to prevent any member from trading, as this would be a direct breach of anti-trust/monopoly, and so illegal for IAATO to do it. However, after 25 years of IAATO's work, there has been only one instance where a company was put on probation (Kim Crosbie, personal communication, 14 March, 2016).

It was suggested that an alternative to enhance enforcement of existing regulation could be to introduce "sponsoring states" for tourism operations, a concept formally adopted by ATCPs within the now expired 1988 Convention on the Regulation of Antarctic Mineral Resource Activities. In the tourism context, this would mean that Treaty Parties that have "a genuine and substantial link" with a tour operator could sponsor him/her. Such a concept would imply that a sponsoring state could be liable for the actions of the sponsored tour operator (Jabour 2014). However, how in practice this concept would work is still uncertain.

A number of mechanisms–either formal or not—that have been used by Treaty Parties and IAATO to monitor the level of compliance of existing legislation from tour operators will be outlined in following sections.

3.6 Discussing Tourism Supervision in the Context of the ATCM

Several papers on the issue of non-Party flags and Port state control, in relation to Antarctic ship borne tourism, have been introduced to the ATCMs since 1996.[12] These documents led to debates in the Plenary, which seemed to reveal that ATCPs were rather unfamiliar with these issues, and in consequence, preferred to adopt conservative positions. For instance, at ATCM XX (Utrecht 1996) some delegations indicated that "the possibility that, in principle, some form of harmonized inspection by Port States might serve a useful purpose, should not be excluded". However, the Meeting agreed "that this issue raised a great many intricate and sensitive questions, and further consideration of this subject could be undertaken only after the Governments of all Consultative Parties would have had the occasion to study this issue in depth" (para 93, Final Report XX ATCM, Utrecht 1996). A year later, "concern was expressed by some delegations about the legality of

[12]These include Chile (1996), The Netherlands (1996), Germany (1997), the United Kingdom (1997), ASOC (2002, 2003) and New Zealand (2007, 2008, 2010).

asserting jurisdiction over future acts of foreign expeditions outside territorial waters". On that occasion, while one Party expressed "that proposals in this context–going beyond the limits of international law, such as departure state jurisdiction—might endanger the delicate balance provided by UNCLOS", another one noted that "legislation of some Parties embodied departure state jurisdiction" (Final Report XXI ATCM, Christchurch 1997 paragraph 14–15).

Over time, Treaty Parties seemed to have softened their positions. By ATCM XXV, a number of countries pointed out that "Port State Control could help to ensure consistency of standards for tourists and other vessels departing for Antarctica" (par. 120 Final Report XXV ATCM, Warsaw 2002), and by 2010 the Meeting eventually adopted a resolution on the enhancement of Port State Control (Resolution 7, 2010 *Enhancement of port State control for passenger vessels bound for the Antarctic Treaty area*). Such a resolution will be further discussed in following sections.

The first time issues specifically related to supervision of Antarctic tourism operations were tackled at the ATCM was in 2010, when Argentina presented Working Paper 48 "Supervision of Antarctic tourism", on the basis of comments made at ATCM XXXII (Baltimore 2009) by different ATCPs and IAATO regarding possible mechanisms for ensuring more appropriate supervision of tourism on-board cruise ships (Argentina 2010). This presentation aimed to discuss possible additional tools, or changes to existing tools, that would serve to ensure better supervision of the management of Antarctic tourism on-board cruise ships.

Subsequently, an Intersessional Contact Group (ICG) was established in the framework of the ATCM. The ICG operated for 2 years and its work eventually resulted in Resolution 9 (2012) *The Assessment of Land-Based Expeditionary Activities*, a checklist to assist inspection/observation schemes of in-field conduct of visitor's activities under Article VII of the Antarctic Treaty and Article 14 of the Madrid Protocol.[13] The work of this ICG also gathered and analyzed information about existing and previous tourism and non-governmental "observer programmes" in the Antarctic Treaty area, and examined the need and identify potential options to enhance the observation of tourism and non-governmental activities in Antarctica (Argentina 2010, 2011, 2012). Different suggestions for enhancing and increasing the number of tourism inspections/observations were also presented during the ICG discussions, some of which will be further considered in this article.

The analysis of the ATCM debates on the supervision of Antarctic shipborne tourism seems to indicate that ATCPs have not addressed these issues in a consistent manner. It would rather seem that they have brought these matters up in a spasmodic way. That is, without taking adequate stock of what had been discussed and agreed before, without a comprehensive view of the whole issue.

---

[13]This checklist supplemented other existing Checklists agreed at ATCMs by Resolution 5, 1995; Resolution 4, 2008; and Resolution 3, 2010, which cover operations of Antarctic stations, ships operating in Antarctic waters and protected areas, respectively.

3.7 Existing Mechanisms to Supervise Antarctic ShipBorne Tourism

The different supervision mechanisms that have been (or are regularly) carried out by ATCPs and IAATO include formal inspections and non-formal observation schemes,[14] as well as Flag and States Port Control. While the former are implemented on-board cruise ships en route to Antarctica or at Antarctic landing points, Flag and State's Port Control are usually exercised outside Antarctica, before vessel departure.

3.7.1 Inspections

An inspection is a formal mechanism specified in Article VII of the Treaty (and article 14 of the Protocol) aimed to ensure the observance of the Antarctic Treaty regulations in place. According to the Treaty, each Contracting Party has the right to designate observers to carry out inspections of any station, installation and equipment in Antarctica, including all ships and aircraft operating in the area at points of disembarkment. In addition, Article 14 of the Protocol presents the promotion of environmental issues as an additional objective of an inspection.

However, inspections have traditionally focused on research stations, rather than on tourism vessels. Since the signature of the Madrid Protocol in 1991 until 2014/15, only 6 out of 30 inspections included observation of tourist cruise ships, totaling 25 tour ships inspected.[15] (ATS 2014 web site, accessed 16 March, 2016). More recently, the ATCM discussed the appropriateness of such inspections with respect to obligations under Art. VII (3) of the Antarctic Treaty, which states that "All areas of Antarctica, including… all ships and aircraft at points of discharging or embarking cargoes or personnel in Antarctica, shall be open at all times to inspection… by any designated observers". (XXXVIII ATCM, para 84–87). This discussion focused on yachts rather than other forms of shipborne tourism, although conceptually the same issues apply.

In addition, it has been pointed out that significant differences seemed to exist between inspection reports discussing stations (and other facilities operated by National Antarctic Programs) and those discussing tourist vessels. Some inspections of cruise ships would appear to be rather general, and consequently less critical, when compared to those of Antarctic stations and related facilities. That would result in—or perhaps be caused by—Treaty Parties having much lesser

[14]The technical distinction between inspections and observation schemes is just a procedural one, as the main differences between them lie in the formal procedure of observers' designation set up in article VII of the Treaty and in the reporting procedures included in article 14 of the Madrid Protocol. Apart from these, same procedures should be applied by any inspection or observation team when dealing with obtaining information from appropriate sources.

[15]In the same period, over160 visits were carried out by inspection teams to Antarctic stations.

knowledge of tourism activities, when compared to station activities. However, since tourism is now a major Antarctic activity in terms of the number of people, ships and sites involved, it would make sense that tourism became the focus of inspections to a greater degree than hitherto (ASOC 2010). Deploying inspection teams to undertake tourism-only inspections would be an option worth exploring. Such a practice would provide a greater basis for comparison of tourist activities (i.e. it would allow comparing how different operators do the same thing), and should not conflict or compete with inspections to land facilities.

Tourism-focused inspections would improve the inspection regime to include a broader range of activities, but would not in themselves suffice to supervise ship borne tourism. Taking into account the resources needed to annually deploy a group of observers to Antarctica, and the fact that inspections can only provide a general picture of the vessel inspected (due to the short time available for conducting it and coverage limited to on-board activities) inspections should only be taken as a complementary, but rather limited, way to supervise compliance of tour operators with existing regulation (see e.g. Sánchez and Njaastadt 2014).

Finally, it is worth noting that reports of inspection teams are formally presented at ATCMs, providing the Meeting the opportunity to be informed and discuss their results. As such reports are made publicly available after the closure of ATCMs, they become an invaluable source for research studies on *inter alia* the level of compliance of Antarctic tourism related regulation by the tour industry.

3.7.2 Official Observer Programmes On-board of Cruise Ships

So far, official observer programmes on-board of Antarctic cruise ships have been scarcely implemented by Treaty Parties. New Zealand is the only ATCP which requires a government observer to be carried on all tourist vessels departing its ports for Antarctica, under its legislation implementing the Madrid Protocol (ATS 2010: p. 235). Argentina, Australia and Norway have occasionally deployed observers on tourist expeditions to the Antarctic as well to audit compliance with regulations in place under the Treaty, the Madrid Protocol and related instruments agreed at ATCMs.

3.7.3 IAATO Observer Programmes On-board of Cruise Ships

Tour operators who want to become IAATO Members are required to carry an independent IAATO observer on their vessels, whose report is submitted to the Members. However, there is no formal commitment for additional observations at periodic intervals during the operators Membership period

It is only at specific instances, on a case-by-case basis, that an operator is subsequently requested by IAATO to carry an observer.[16] In addition, if a Member does not operate for three consecutive seasons, IAATO requires this Member to go through the full membership reapplication process with a desk top review and then field observation before being put up for vote again to become a Member. Members can also volunteer for an observer at any time. By doing so, they benefit from having an opportunity to discuss their operation with an experienced third party. This can be helpful if Members want to do a "health check" of their operations, or if they have amended their operations significantly. While the majority of IAATO observers are recruited independently, IAATO is gradually moving to incorporate its own personnel, particularly for the voluntary observations and/or for operators introducing completely new activities, for which IAATO Secretariat needs to understand them better (Kim Crosbie, personal communication, 14 March, 2016).

IAATO has recognized that, although this system has proved broadly effective, there was scope for it to be strengthened. Among the issues IAATO considered useful to discuss to enhance its Observer Scheme in the future, are matters connected to confidentiality, the frequency at which the observation review process should take place, the likely development of a dedicated, accordingly trained, team of observers[17] and the financial responsibilities of the observer activities, among others (IAATO 2011).

Circulation of information on observer programmes is not a requirement under the Antarctic Treaty Information Exchange System. And only occasionally, have Treaty Parties and IAATO reported such activities within ATCMs. As the reports of such observation activities have rarely (or never) been presented as formal documents to ATCMs—unlike the reports of inspection teams—access to their results to ATCPs and the general public is therefore very limited, or negligible. Consequently, this does not allow determining how proactive they are in conducting such activities.

In this sense, during the debates on the ICG on Supervision of Antarctic Tourism referred to above, some Treaty Parties pointed out that placing ATCP observers on tour ships (in cooperation with the operator) under their supervision was a very positive experience, since such programs allowed them to fully assess tourist activities during a complete trip (covering both on-board and in-field activities); provided in situ management feedback and promoted mutual understanding between the tour operator and the National Antarctic Program. IAATO noted that they believe strongly in the observer process (Argentina 2011).

It has been suggested that the presence of independent observers on-board of IAATO vessels in order to audit operator compliance would be desirable, as a likely means of improving the enforcement of current regulatory mechanisms (Haase 2008; Liggett et al. 2011; Jabour 2014).

---

[16]For instance, this happened to GAP Expeditions after the loss of the MS Explorer (IAATO 2011).

[17]IAATO recruits observers on the basis of his/her related experience and understanding of field operations, and avoiding any conflict of interest. They are to perform the observer activities at their own costs, except for travel expenses, which are covered by the company being observed.

3.7.4 Observer Programmes at Landing Sites

In 2006 a team of representatives from five ATCPs and the IAATO conducted an on-site review of ten sites for which "Site Guidelines for Visitors" had been prepared in the north and west of the Antarctic Peninsula. The aim of the visits was to review *in situ* the management provisions of such Guidelines and to provide up-to-date information on these sites (United Kingdom et al. 2006). A similar exercise was replicated in 2013 (United Kingdom et al. 2013). These observer programs stand perhaps as the only policy-oriented exercises ever done with the purpose of supervising tour activities at Antarctic landing sites.

Within this framework, observer teams (although not formally designated as an inspection team) were deployed in sites highly visited by tourism. Collecting representative data in this manner would be, however, highly resource demanding, especially considering the fact that for effective monitoring, one would require a data series covering at least 10 years for each site (Argentina 2011).

Despite the fact that the Antarctic Treaty stipulates that an inspection team can conduct its activities not only on-board the cruise ships, but also at landing points, there has been limited use of a scheme as such, with the exception of the above mentioned exercises.[18]

3.7.5 Flag and Port States Control

Concerning the Flag State Control, it is important to note that more than 50 % passenger vessels operating in the Antarctic Treaty area are not flagged to States which are Parties to the Antarctic Treaty or to its Protocol on Environmental Protection[19] (Jabour 2014). The high proportion of vessels in the tourism trade flagged to non-Parties has already been considered an issue of significant concern for Treaty Parties (ATCM XXX Final Report, page 40 para 178)

In particular, within the framework of ATCMs, it has been suggested that 'non-Party' vessels have implications for the effectiveness of the Antarctic Treaty System, mainly due to the facts that (a) there is no obligation for non-Party Flag States to provide emergency response action under Article 15 of the Protocol and (b) in the case of a non-Party vessel, there is no obligation to allow for an inspection under Article VII of the Antarctic Treaty and under Article 14 of the Protocol. In

[18]Outside the formal structure of the ATS, the organization Oceanites, with the support of IAATO members, has been conducting site inventories at tourism landing sites, which focus primarily in flora and fauna, and apparent impacts, rather than on human activity itself. A somewhat similar activity has been carried out by the New Zealand Antarctic Program (New Zealand 2006, 2009b) in site assessments known as Visitor Site Assessment Scheme (VISTA). However, VISTA is focused on environmental impact assessment rather than the supervision of tourism per se.

[19]Notwithstanding, all of the currently known tourism operators are located in a State party to the Antarctic Treaty and the Protocol on Environmental Protection (Vöneky 2016).

addition, according to Annex VI of the Protocol, which is yet to enter into force, responsibility for response action lies with the "operator" (those organizing or conducting Antarctic activities), who are required to take prompt, effective response action when environmental emergencies arise from their activities. If an operator fails to take prompt, effective response action, a Party may do so, and the defaulting operator will have to reimburse the Party's costs. In the case of shipborne tourism it is likely that in most instances the "operator" for the purposes of Annex VI will be a tour company which has chartered a vessel for an expedition to the Antarctic Treaty area, and completed the Protocol's environmental impact assessment procedures with one of the Treaty Parties (New Zealand 2010). This, however, should not override the Flag State's ultimate responsibility for the vessel (New Zealand 2009a).

In regards to the Port State Control of passenger vessels bound for the Antarctic Treaty area, the existing regime, which is exercised by national maritime authorities, only includes the last port prior to departure for Antarctica, ultimately leaving the responsibility of inspecting vessels to a very limited number of port States. As the existing international maritime regulation in areas beyond port or coastal state jurisdiction is the primary responsibility of the Flag State, the combination of Flag State and focused and consistent existing Port State Control scheme measures to enforce IMO instruments relevant to polar waters should be sufficient to increase ship safety and protection of Antarctic waters from pollution from ships.

The fact that in 2010 the ATCM adopted Resolution 7 (*The Enhancement of Port State Control for Passenger Vessels Bound for the Antarctic Treaty Area*), which recommends that the ATCPs proactively apply, through their national maritime authorities, the existing regime of port State control to passenger vessels bound for the Antarctic Treaty area, was probably a reflection from ATCPs that more needed to be done in terms of port State control. The text eventually adopted is, however, very open and leaves room to accommodate concerns expressed by Treaty Parties in previous ATCMs.

During the debates on the adoption of such Resolution, several ATCPs agreed that its objective was to indicate that it was important for Treaty Parties to give priority to passenger vessels bound for the Antarctic Treaty area when conducting port State control. It was also suggested that a high priority status for passenger ships could help to ensure they were inspected at regular intervals by at least one Party, using common inspection guidelines. It has also been suggested that the ATCPs encourage the secretariats of the various existing Port State Control memoranda of understanding to share information on inspections of vessels departing to Antarctica[20](ATS 2010: p. 308; 314–315). The traditional southern hemisphere countries of vessel departure are all ATCPs, are part of one or more regional

[20]Such Memoranda of Understanding on PSC include: Europe and the north Atlantic (ParisMoU); Asia and the Pacific (TokyoMoU); Latin America (Acuerdo de Viña del Mar); Caribbean (Caribbean MoU); West and Central Africa (AbujaMoU); the Black Sea region (Black Sea MoU); the Mediterranean (Mediterranean MoU); the Indian Ocean (Indian Ocean MoU); and the RiyadhMoU.

memoranda on port State control and have all ratified SOLAS 74 and Annexes I and II of MARPOL 73/78[21](New Zealand 2010). In this context, it has been pointed out that, in order to achieve full effectiveness, PSC inspections to vessels departing to Antarctica should be harmonized and standardized in all port states, and all ATPs that would participate in port state jurisdiction should be adequately equipped and manned to provide the required levels of inspection (Haase 2008). However, to what degree are such port states prepared to assume greater jurisdictional responsibility in regulating and managing Antarctic shipborne tourism is a question that needs to be addressed before elaborating further proposals.

Unfortunately, resolutions of that kind are seldom followed up adequately, so as to monitor how ATCPs have responded to the norm. If the situation remained unchanged, at best a new resolution would come up in the future, reminding ATCPs of Resolution 7, 2010.[22] Treaty Parties should also be urged to periodically inform the Meeting how such resolutions are being implemented, perhaps by including a specific requirement in the Antarctic Treaty Information Exchange System.

3.8 Enhancing the Supervision of Antarctic Tourism

The discussion above suggests that the mechanisms in place to supervise Antarctic shipborne tourism are limited at present. To the extent that there is a regular ground supervision of tourism this lies primarily on the hands of the tourism industry itself, which is useful but not ideal. Therefore, likely additional mechanisms to enhance the observation of tourism and non-governmental activities in Antarctica could include the following:

1. Arrangements between IAATO members and individual National Antarctic Programs (NAP), which could include "observation by invitation" schemes. This could occur under two complementary scenarios:

    - Those ATCPs interested in deploying observers on an IAATO tour ship would so inform IAATO, which would in turn inform its members of such interest. Those IAATO members who volunteered to carry observers aboard their tour ships would then coordinate with IAATO and the Party how these activities could take place (date, work conditions, confidentiality, travel expenses, insurance, etc.).

[21]The need for the agreement of an Antarctic Port State Control Memorandum of Understanding has been also proposed, though no consensus has yet been found on this idea, as some think the existing MoUs are sufficient to deal with Antarctic port state control.

[22]Similarly, Resolution 7 (2014) was adopted to remind Parties that Measure 4(2004) "Insurance and Contingency Planning for Tourism and NGO activities in the AT Area" should indeed be ratified after 10 years of having been agreed, without having entered into force yet.

- Those IAATO members interested in carrying observers from NAPs would so inform IAATO, which would in turn inform NAPs of such interest. Those NAPs which volunteered to carry out observation activities on IAATO tour ships would then coordinate with IAATO and the tour operator how these activities could take place (date, work conditions, confidentiality, insurance, etc.)

Formal Agreements (e.g. Memoranda of Understanding) or other agreements may possibly be needed between IAATO and ATCPs interested in deploying observers on an IAATO tour ship to facilitate the conduct of these programs. Additional agreements between gateway ports countries and ATCPs interested in deploying observers on an IAATO tour ship could also be necessary.

2. Bilateral agreements between NAPs, which would in turn may comprise:

- Mutual exchange of observers, which would consist of allowing observers from NAP X to board ships flying a flag of NAP Z (or whose tour operator is registered in NAP Z), and observers from NAP Z to board ships flying a flag of NAP X (or whose tour operator is registered in NAP X). Again, formal Agreements (e.g. Memoranda of Understanding) or other agreements between ATCPs interested in carrying out this mutual exchange programs would contribute to officially shape these activities.
- Invitation to join observation schemes in place. As is usually being done in recent inspections, a Party conducting observation activities on vessels flying its flag (or whose tour operator is legally registered in its territory), could invite another Party to join its observation team.

In this respect, it was noted that no impediments exist either for any Treaty Party (or group of Treaty Parties) choosing to conduct inspection/observation activities in these ways, or for tour operators and ATCPs to make arrangements between themselves for the placement of observers on-board tour vessels. These approaches would seem to require no ATCM endorsement, as Treaty Parties can pursue them if they regard them as a priority (Argentina 2011).

3. Observer programmes at fixed landing sites

An alternative to the observer programmes on-board vessels described in the previous section would include the designation of personnel who were regularly undertaking National Antarctic Programs's activities on fixed sites (stations, huts, refuges, field camps) where field tourist activities usually take place, as observers according to article VII of the Treaty. In that way, human resources would be optimized: observers would not go to Antarctica just to undertake inspections. They would do so occasionally (upon tourist arrival), as part of other duties. Although information gathered in this way would be restricted to some sites only (e.g. during the 2013/14 season, tourist vessels visited a total of 20 Antarctic stations plus some Antarctic huts and refuges), this mechanism would at least help to cover part of the gap (each season, some 80 sites receive regular visits[23]). It was suggested, however, that using station personnel to observe/inspect tourist activities/operations may affect the work of the station since this personnel may be already fully engaged in their dedicated roles (Argentina 2011). Such an assumption is, however, heavily dependent on the site's level of visitation.

An enhanced implementation of observers' activities would result in mutual learning experiences for those being observed and for the ATCPs conducting such programs. Treaty Parties would become much more familiar with the practical, in situ management of Antarctic shipborne tourism (which would help them to understand, and thus to adequately regulate it), while tour operators would become much more familiar with the practical implementation and meaning of regulating instruments in force. This would be a win–win situation, where the general result would be a higher level of confidence between ATCPs and tour operators, and higher standards for both regulation and management of Antarctic ship borne tourism. If these in situ observation activities are multiplied, the whole Antarctic Treaty System will gradually build up a reliable database on these activities. Again, it would be a practice that would provide a greater basis for comparison and understanding.

3.9 Conclusions

Antarctic shipborne tourism is regulated through a number of mandatory and non-mandatory provisions emerging from the ATCM, complemented by industry self-regulation applicable to IAATO members. At present these regulations cover broadly most of the situations that could lead to impacts to navigation safety and to the Antarctic environment. However, the enforcement of such regulations in the Antarctic context constitutes a challenge for ATCPs mainly due to the lack of effective enforcement capabilities in the continent. In addition, tourism is a dynamic industry and new developments may require new regulation in the future. This may include changes in fly–sail tourism promoted by e.g. airstrip development, ability of tour ships for refuelilng at sea, and/or the establishment of base camps in coastal areas (distinct from air supported, land-based camps).

Inspections of shipborne tourism under Article VII of the Antarctic Treaty and Article 14 of the Protocol have so far been a rather limited option to effectively monitor the tour operators' level of compliance. Information provided by ATCPs (through the AT Information Exchange System and documents presented to ATCMs) and by IAATO (through papers presented to ATCMs) on the conduct of inspection/observation schemes has also been limited, which suggests that such activities might be still insufficient to ensure a proper supervision of the existing applicable regulation in practice.

[23]For the 2013/14 season, 75 Antarctic sites—out of 218—have received more than 1000 visitors.

In addition, doubts as to the effectiveness on the Flag State regime, added to the concerns expressed by ATCPs over the full applicability of Port State Control in the Antarctic context, also contribute to add uncertainties on the likely means of achieving a stricter compliance of the corpus of regulations in place.

Since the 1990s, coincident with the growth of Antarctic shipborne tourism, ATCMs have addressed the issue of Supervision of Antarctic tourism through intersessional debates; have expressed concern over vessels flagged to non-Party countries; and they have even adopted resolutions where Treaty Parties have recognized that there was a need to proactively develop regulations relating to tourism activities and to enhance Port State Control, among other related issues. All these elements together seem to indicate that ATCPs are currently aware that issues connected to supervision of Antarctic shipborne tourism need further work.

However, recognizing the existence of a problem does not constitute the solution itself. The adoption of agreed tools with specific references to shipborne activities (including *inter alia* inspection checklists, Antarctic protected areas' management plans and EIAs conducted as part of the authorization process) are useful to guide such inspection/observers programs and should be taken as a positive step that should be duly recognized.

Looking ahead into the future, further debates within the ATCM—preferably through intersessional discussions—on the likely approaches to enhance Port State Control in the Antarctic context, aimed to complement the current Flag state regime, would be a valuable means of improving supervision of shipborne tourism in Antarctica. In-depth analysis of the correlation between pre-season and annual cruise ship information provided by ATCPs to the Antarctic Treaty Information Exchange System would also be an invaluable input for further studies. Finally, Treaty Parties and IAATO should be encouraged to enhance, broadening and formalize, in close cooperation and through innovative measures, their supervision activities (i.e. observer programs) of Antarctic tour ship operations, and to continuously report to ATCMs on the activities carried out with that purpose.

Overall, Treaty Parties should be more proactive, innovative and, above all, more cooperative to collectively address the question of how to enhance the supervision of Antarctic shipborne tourism. Countries who are directly linked to Antarctic tourism would naturally need to be the most proactive to conduct proper supervision of such activities. These would include those ATCPs which have Antarctic tour operators registered in their territories; those whose flagged vessels usually operates in Antarctic waters; and those whose port facilities are used as tourist gateways to Antarctica. However, other ATCPs would need to be involved as well in the supervision of tourism.

These matters pose a challenge that, if consistently and collectively addressed by Treaty Parties with the support of the tour industry, would allow for a proper assessment and enhancement of the compliance degree of existing regulations on Antarctic tourism ship operations, and hence, for a higher level of transparency of such activities. Lack of supervision of Antarctic tourism activities may result in a gap between the theory—what the regulations are and how these are implemented and reported at the ATCM—and the practice of tourism on the ground.

3.10 Disclaimer

The views and opinions expressed in this article are those of the authors and do not reflect nor compromises the positions of the institutions they represent.

Acknowledgments The authors thank Machiel Lamers, Máximo Gowland, Kim Crosbie and José María Acero, for their valuable comments on this chapter.

References

Amelung, B., & Lamers, M. (2006). Scenario development for antarctic tourism: Exploring the uncertainties. *Polarforschung, 75*(2/3), 133–139.

Antarctic Treaty Secretariat. (2014). *ATCM Multi-Year Strategic Work Plan: Summary of the ATCM discussions and decisions on land-based and adventure tourism*, presented as Secretariat Paper 9 to ATCM XXXVII (Brasilia, 2014).

Argentina. (2010). *Supervision of Antarctic tourism*, presented as Working Paper 48 to ATCM XXXIII (Punta del Este, 2010).

Argentina. (2011). *Report of the intersessional contact group on supervision of Antarctic tourism* presented as working paper 48 to ATCM XXXIV (Buenos Aires, 2011).

Argentina. (2012). *Final report of the intersessional contact group on supervision of Antarctic tourism* presented as working paper 43 to ATCM XXXV (Hobart, 2012).

ASOC. (2002). Port state jurisdiction: An appropriate international law mechanism to regulate vessels engaged. In *Antarctic tourism,* presented as IP 063 to ATCM XXV (Warsaw, 2002).

ASOC. (2003). *Port state control: An update on international law approaches to regulate vessels engaged.* In *Antarctic non-governmental activities*, presented as IP 044 to ATCM XXVI (Madrid, 2003).

ASOC. (2010). *Antarctic shipborne tourism and inspections under article VII of the Antarctic treaty and article 14 of the protocol on environmental protection.* Presented as IP 82 to ATCM XXXIII (Punta del Este, 2010).

Bastmeijer, C., & Roura, R. (2004). Regulating Antarctic tourism and the precautionary principle. *American Journal of International Law, 98*(4), 763–781.

ATS. (2010). *Final report of the thirty-third Antarctic Treaty Consultative Meeting*. Buenos Aires: Secretariat of the Antarctic Treaty. Retrieved from: http://www.ats.aq/documents/ATCM33/fr/ATCM33_fr001_e.pdf (Punta del Este, 2010).

ATS. (2013). *Final report of the thirty-sixth Antarctic Treaty Consultative Meeting* (Vol. 1). Buenos Aires: Secretariat of the Antarctic Treaty. Retrieved from: http://www.ats.aq/documents/ATCM36/fr/ATCM36_fr001_e.pdf (Brussels, 2013).

Bastmeijer, K. & Roura, R. M. (2008). Environmental impact assessment in Antarctica. In K. Bastmeijer & T. Koivurova (Eds.), *Theory and practice of transboundary environmental impact assessment, monographs series on legal aspects of sustainable development* (pp. 175–219). Leiden: Brill/Martinus Nijhof Publishers.

Cessford, G.R. & Dingwall, P.R. (1996). *Tourist visitors and their experiences at New Zealand subantarctic islands* (Vol. 1). Wellington, New Zealand: Department of Conservation (Science & Research series, 0113–3713 ; no. 96).

Chile. (1996). *Enforcement of international maritime conventions and domestic standards on the inspection of ships that operate in the Antarctic* presented as WP 17 to XX ATCM (Utrecht, 1996).

Germany. (1997). *Inspection of ships by port states* presented as WP 16 to XXI ATCM (Christchurch, 1997).

Haase, D. (2008). *Tourism in the Antarctic. Modi operandi and regulatory effectiveness.* PhD thesis. Gateway Antarctica: University of Canterbury.

Hall, C. M. (1992). Tourism in Antarctica: Activities, Impacts, and Management. *Journal of Travel Research, 30*(4), 2–9. doi: 10.1177/004728759203000401

Hemmings, A. D., & Roura, R. M. (2003). A square peg in a round hole: Fitting impact assessment under the Antarctic environmental protocol to Antarctic tourism. *Impact Assessment and Project Appraisal., 21*(1), 13–24.

IAATO. (2011). *Towards An IAATO enhanced observer scheme*, presented as IP 107 to ATCM XXXIV (Buenos Aires, 2011).

Jabour, J. (2011). Would you like ice with that? Antarctic tourism and climate change. In A. Jones & M. Phillips (Eds.), *Disappearing destinations: Climate change and the future challenges for coastal tourism*, pp. 177–190. UK: CABI.

Jabour, J. (2014). *Strategic management and regulation of Antarctic tourism.* In: T. Tin, D. Liggett, P. Maher & M. Lamers (Eds.), *Antarctic Futures: Human Engagement with the Antarctic Environment* (pp. 273–286). Springer.

Lamers, M., Haase, D., & Amelung, B. (2008). Facing the elements: Analysing trends in Antarctic tourism. *Tourism Review, 63*(1), 15–27. doi:10.1108/16605370810861017.

Lamers, M., Liggett, D., & Amelung, B. (2012). Strategic challenges of tourism development and governance in Antarctica: Taking stock and moving forward. *Polar Research,*. doi:10.3402/polar.v31i0.17219.

Liggett, D, McIntosh, A., Thompson, A., Gilbert, N., Storey, B. (2011). *From frozen continent to tourism hotspot? Five decades of Antarctic tourism development and management, and a glimpse into the future*. Tourism Management 32 (2011) 357e366. doi:10.1016/j.tourman.2010.03.005

Molenaar, E. J. (2005). Sea-borne tourism in Antarctica: Avenues for further intergovernmental regulation. *International Journal for Marine and Coastal Law, 20*(2), 247–295.

New Zealand. (2006). *An update on the antarctic visitor site assessment scheme*: VISTA presented as 011 to XXIX ATCM (Edimburgh, 2006).

New Zealand. (2007). *Tourist vessels flagged to non Parties; Implications for the effectiveness of the Antarctic treaty system* presented as WP 14 to XXX ATCM (New Delhi, 2007).

New Zealand. (2008). *Implications of tourist vessels flagged to non-Parties for the effectiveness of the Antarctic treaty system* presented as WP 22 to XXXI ATCM (Kyiv, 2008).

New Zealand. (2009a). *A proposal to enhance port state control for tourist vessels departing to Antarctica* presented as WP 7 to ATME on Maritime Tourism (Wellington, 2009).

New Zealand. (2009b). *Joint VISTA-oceanites antarctic project* presented as IP037 to XXXII ATCM (Baltimore, 2009).

New Zealand. (2010). *The enhancement of port state control for passenger ships departing to Antarctica* presented as WP 37 to XXXIII ATCM (Punta del Este, 2010).

Orrego Vicuña, F. (2000). Port state jurisdiction in Antarctica: A new approach to inspection, control and enforcement. In: D. Vidas (Ed.), *Implementing the Environmental Protection Regime for the Antarctic* (pp. 45–49).

Plaza, F. (1999). The future for flag state implementation and port state control. In: M. H. Nordquistand & J. N. Moore (Eds.), *Center for Oceans Law and Policy: Current Maritime Issues and the International Maritime Organization* (p. 199, 201).

Sánchez, R. A. & Njaastad, B. (2014). Future challenges in environmental management of National Antarctic Programs. In: T. Tin, D. Liggett, P. Maher, & M. Lamers (Eds.), *Antarctic Futures: Human Engagement with the Antarctic Environment.* Springer.

Scott, S. (2001). *How cautious is precautious? Antarctic tourism and the precautionary principle*, International and Comparative Law Quarterly (Vol. 50, October, pp. 963–71).

The Netherlands. (1996). *Inspection of ships in gateway ports to Antarctica, on the basis of Marpol 73/78, and in Antarctic ports under the Environmental Protocol (Annex IV) to the Antarctic Treaty* presented asWP 9 to XX ATCM (Utrecht, 1996).

The Netherlands. (2014). *Report of the informal contact group on the increasing diversity of tourism and other non-governmental activities in Antarctica* presented as WP 47 to ATCM XXXVII (Brasilia, 2014).

United Kingdom. (1997). *Enhancing compliance with the protocol: departure state jurisdiction* presented asWP 22 to XXI ATCM (Christchurch, 1997).

United Kingdom, Argentina, Australia, Norway and United States. (2006). *Policy Issues Arising from On-Site Review of Guidelines for Visitor Sites in the Antarctic Peninsula presented* as WP 2 to ATCM XXXI (Edinburgh, 2006).

United Kingdom, Argentina, Australia and United States. (2013). *Policy Issues Arising from the 2013 On-Site Review of Guidelines for Visitor Sites in the Antarctic Peninsula*as WP 15 to ATCM XXXI (Brussels, 2013).

Valenzuela, M. (1999). Enforcing rules against vessel-source degradation of the marine environment: Coastal, flag and portstate jurisdiction. In: D. Vidas & W. Østreng (Eds.), *Order for the Oceans at the Turn of the Century* (p. 485, 497).

Vöneky, S. (2016) *Analysis and enhancement of the legal framework—The need for action in the light of current developments of Antarctic tourism.* Environmental Research of the Federal Ministry for the Environment, Nature Conservation, Building and Nuclear Safety. On behalf of the German Environment Agency. Retrieved January 2016, from http://www.umweltbundesamt.de/publikationen/analysis-enhancement-of-the-legal-framework-the. ISSN 1862-4804, Dessau-Roßlau.

Chapter 4
Adventure Tourism Poses New Challenges for the International Association of Antarctica Tour Operators (IAATO)

Monika Schillat

Abstract The chapter focuses on how the IAATO (International Association of Antarctica Tour Operators), member organization, which aims to advocate and promote the practice of safe and environmentally responsible travel to the Antarctic has dealt over the last 25 years with visitor guidance and the management of the tour operators. It also discusses how the apparent expansion of new activities, such as adventure tourism proposals, carried out in Antarctica, pose new challenges to both Treaty Parties and IAATO, keeping in mind aspects of risk management and environmental considerations. We aim to analyze IAATO's strategies towards these new activities.

Keyword IAATO · Visitor guidance · Management of tour operators · Expansion of new activities · Adventure tourism

Tourism in Antarctica has rapidly increased in recent decades. In terms of technical practice and visitor guidance, the management of the tour operators is seemingly determining whether tourism has a negative or positive impact on the environment. But these days, there is not only an apparent expansion of new activities carried out in Antarctica but also more and more non-regulated tourism activities. Treaty Parties focus on the perceived challenges posed by such adventure tourism. The IAATO (International Association of Antarctica Tour Operators), member organization, which aims to advocate and promote the practice of safe and environmentally responsible travel to the Antarctic has to deal with new proposals made by tour operators and travel agents, keeping in mind aspects of risk management and environmental considerations. At the same time, the activity of non-IAATO operators demand attention by both, IAATO and Treaty Parties. In the following, we aim to analyze IAATO's strategies towards these new activities—including both

---

M. Schillat (✉)
Instituto de Ciencias Polares, Ambiente y Recursos Naturales,
Universidad Nacional de Tierra del Fuego, Onas 450, 0910 Ushuaia, Argentina
e-mail: mschillat@untdf.edu.ar

M. Schillat et al., *Tourism in Antarctica*,
SpringerBriefs in Geography, DOI 10.1007/978-3-319-39914-0_4

deep field and those additional or new activities taking place from traditional ship or yacht-based platforms.

Antarctica receives relatively few visitors compared to many other wilderness destinations. Nevertheless, IAATO and the wider Antarctic community work together to understand tourist activities and safeguard the environment from any cumulative impact. Careful management and monitoring of the landing sites seems to be the key. IAATO has actively assisted Treaty Parties to develop sound and practical guidelines for the most-visited sites in Antarctica. Over the past years, IAATO has proposed a number of guidelines and procedures that members should follow, in conjunction to its internal by-laws, some of which have constituted the basis for the Antarctic Treaty System's (ATS) own regulations (Jabour 2014). Until today, there has been virtually no observed impact of more than a minor or transitory nature to the continent. This is largely because a coalition of Antarctic tour operators has worked for over two decades to develop a system of self-regulation and management to protect this last great wilderness for future generations. The members of this unlikely coalition are celebrating this year, their 25th Anniversary and there is a lot to celebrate. Theirs is a success story. Over the last 25 years, IAATO has shown that with careful management it is possible to tread softly on the White Continent. Working together under the umbrella of IAATO, competing tour operators and supporting companies, played an important and significant role in the conservation and management of the planet's last great wilderness.

When the IAATO was founded as a member organization back in 1991, only seven Antarctic tour operators were present. These had been operating expeditions to Antarctica for several years already, when the signing of the Antarctic Environmental Protocol in early 1991 by the Antarctic Treaty nations established rigorous standards beyond those already outlined in the Antarctic Treaty of 1959. These standards designated Antarctica as a "natural reserve dedicated to peace and science." The Protocol provided the framework for the continued protection of the Antarctic environment. By joining together, the companies were able to pool resources and lobby their national governments to develop substantive regulations and guidelines compatible with the best practices that were already being followed in the field.

In the meantime, IAATO has grown to include over 120 member companies from all over the world. The membership consists of approximately 60 operators, who organize and conduct expeditions to Antarctica. Another 60 companies—comprising travel agents, ship agents and conservation groups that work within the Antarctic community—have chosen to join as non-operating members of the Association supporting the mission and vision of the organization. (The membership directory can be found at www.iaato.org).

IAATO is dedicated to facilitating appropriate, safe and environmentally sound private-sector travel to the Antarctic, to represent its members to the Antarctic Treaty Parties, the international conservation community and the public at large. The association has circulated and promoted the General Guidelines for Visitors to the Antarctic as adopted by the Antarctic Treaty Consultative Parties under Resolution 3 (2011) and Guidance for Those Organising and Conducting Tourism and Non-governmental Activities in the Antarctic, as adopted by the Antarctic Treaty Consultative Parties under Recommendation XVIII-1 (1994). Members subscribe to the principle that their planned activities will have no more than a minor or transitory impact on the Antarctic environment. IAATO fosters cooperation among its members and provides a forum for the international, private-sector travel industry to share expertise, opinions and best practices. This also includes employment by the membership of the best qualified staff and field personnel through continued training and education; and to encourage and develop international acceptance of evaluation, certification and accreditation programs for Antarctic personnel. Since 2010, IAATO has represented all passenger vessels operating in Antarctic waters under the International Convention for the Safety of Life at Sea (SOLAS).

At the same time, another important objective for IAATO is to enhance public awareness and concern for the conservation of the Antarctic environment and its associated ecosystems and to better inform the media, governments and environmental organization about private-sector travel to these regions. IAATO tries to create a corps of ambassadors for the continued protection of Antarctica by offering the opportunity to experience the continent first hand. Already back in 1966, the concept of expedition cruises coupled with education as a major theme, began with Lars-Eric Lindblad, who sailed an ice-strengthened tourist vessel to Antarctica during austral summers. Since then, the idea of educational voyages has been upheld by IAATO members. Powell et al. (2008) conducted two complementary studies to research both the immediate and longer-term influence of Antarctic cruise tourism experiences on participant's knowledge of Antarctica, attitudes toward management issues facing the Antarctic region, and environmental behaviours as well as future intentions. In addition, the study investigated tourist's attitudes toward visitor guidelines. The results suggest that Antarctica nature-based tourism operators have the potential to provide experiences that educate the public to the importance of Antarctica. The authors agree that the mission of creating ambassadors has been a success.

Recognition comes also from other parties. In 2016 IAATO Executive Director Dr. Kim Crosbie and IAATO member Jerome Poncet of Golden Fleece Expeditions have been awarded the Queen's Polar Medal. The first has been recognized for her outstanding contributions to the knowledge of Polar visitor management, the second for his pioneering efforts in supplying logistics in support of Polar science and wildlife documentaries for over 40 years. (https://polar-news.com/antarctic/society/151-queen-s-polar-medal-awarded-to-iaato-members) HM The Queen awards the Polar Medal to those who have personally made conspicuous contributions to the knowledge of the Polar Regions and/or have provided outstanding service in support of gaining such knowledge.

At the same time, IAATO has grown to become a valued partner, when it comes to environmental monitoring and providing details about tourism in Antarctica. IAATO provides ATCM and the Committee for Environmental Protection (CEP) with detailed information on member activities in Antarctica and works collaboratively with scientific institutions particularly on long term environmental monitoring and educational outreach. In addition to being an invited expert of the ATCM, IAATO also actively works with the CEP, when it comes to consider the effective implementation of the Protocol on Environmental Protection. IAATO has initiated a large number of operating procedures, which have fed into the CEP′s work. The organization also has strong relationships with the five Maritime Rescue Coordination Centres (MRCCs) and Air Rescue Coordination Centres (ARCC) with responsibility in Antarctica. IAATO also participates as an observer at the annual Council of Managers of National Antarctic Programmes (COMNAP) and increasingly IAATO members are contracted by some national operators to assist with their logistic support for scientific programs. Effective tourism management is a collaborative effort amongst all Antarctic stakeholders and overall most might agree that tourism can be sustainable if the strict by-laws that have been compiled by the umbrella organization IAATO are followed by all visitors. IAATO's primary mission is to conduct environmentally responsible, private-sector travel to Antarctica, and educate passengers in the process. Protection of the environment and adherence to Antarctic Treaty regulations and conventions are mandatory, and tour operators have shown to follow them as a dedicated group (Splettstoesser 2000).

But there are concerns about what Antarctic tourism might mean in the future. There are new players around, which are not members of IAATO, including yachts, land-based programs and individual adventure travellers, whose behaviour might reflect badly on the regulated tourism. This aspect however exceeds the present chapter and will therefore not be discussed further. On the other hand, there seems to be an appetite for new forms of adventure travel in Antarctica, which will have to be catered too, even by IAATO members. Tourism as a dynamic phenomenon has the capacity to change, adapt and evolve. At the beginning of the twenty-first century we are witnessing a growing demand from tourists for amazing and bizarre experiences (Tejvir Singh 2004). Some of the tour companies active in Antarctic tourism also have developed their operations to make them more attractive to clients, initiating activities like diving, kayaking, mountaineering, skiing and the like, or introducing itineraries different from those normally conducted.

These special activities in the Antarctic can be defined as any activity beyond passive tourism where visiting Antarctic wildlife, wilderness and historic sites of interest through ship cruising, landings with walking tours, small boat cruising and general observation under guidance from field staff is asked for.

IAATO insists, that prior to offering, or agreeing to support, any activity, the operator has to ensure that it is included in the operator's authorization (Advance Notification and Environmental Impact Assessment EIA). The umbrella organization tries to provide a decision-making framework for assessing a new activity prior to offering, or agreeing to support, the activity. This might help to evaluate special activities new to the operator or to Antarctica. The checklist provided, complements the specific guidelines and Standard Operating Procedures found in the Field Operations Manual. It also keeps in mind, that the new activity should be included in the Operators Standard Operating Procedures and Emergency contingency plans. Consideration should be given if the activity is appropriate for the Antarctic region with regards to safety and the environment and of course that the activity should have no more than a minor or transitory impact. Most importantly and in accordance with IAATO's mission statement, the activity should have an educational element enhancing the visitors' appreciation of the intrinsic values of Antarctica. A risk assessment has to be conducted as well and an overarching emergency contingency plan should be reviewed to consider the new activity and potential problems. IAATO also asks its operators to ensure an alternative activity is available in case the activity is aborted, especially if wildlife is encountered to prevent having more than a minor or transitory impact.

Browsing the Internet in 2014, in search of new adventures planned for Antarctica, gave us a first idea about new challenges for both the ATCM and IAATO. There are quite a few private skiers/bikers/kiters and others out there, who wish to get to the South Pole. In Germany famous parachute artist Klaus Renz was dreaming of a performance in Antarctica. He wanted to jump out of a plane over the ice and land directly at the South Pole,[1] but at the same time he was aware of the restrictions and the fact, that he most likely would not get a permit to do so. Mountaineer Martin Szwed was meanwhile planning his voyage to a base camp on the Antarctic Continent, from where he then would part to climb Mount Vinson. But that did not seem to be enough for the intrepid traveller. The second goal of his voyage was a solo traverse of the continent, breaking the previous record of only 39 days covering 1300 km.[2] Three experienced lifeguards were willing to challenge fate on their voyage to Antarctica. They planned to paddle on boards, relying only on the strength of their arms and nothing else. Their itinerary would take them in December 2014 from Cape Horn at the southernmost tip of South America to The Antarctic Peninsula.[3] And while this possibly would remain a dream, another adventurer, the Dutch Manon Ossevoort was already on her way to Antarctica. After having crossed Africa with a tractor, she now plans on travelling with a similar vehicle from the Russian Nova Station to the South Pole and back again. Her vehicle has already been taken from Cape Town to Antarctica on a Russian transport airplane. Until Christmas she hoped to cover the 4500 km. Ossevoort planned on building a snowman in Antarctica and burying deep inside his belly the messages, with the wishes of those, she has met on her travels through Africa or received via e-mail. The dreams are digitalized and burnt onto a hard disk, which shall remain for 80 years in the frozen wilderness.[4]

In 2016, the results of an Internet search show that neither Klaus Renz nor the intrepid team of lifeguards did succeed in their dreams. Renz at least managed to jump out of a plane over the White Continent, but did not make it to the pole. But Manon Ossevoort arrived with her tractor and a support group of six, at the

[1]Stuttgarter Nachrichten, "Traum vom Sprung über der Antarktis", 27.11.2014.

[2]Hohenzollerische Zeitung, "Auf dem Weg in die Antarktis", 26.11.2014.

[3]Frankfurter Allgemeine Zeitung, "Paddeln von Kap Hoorn zur Antarktis", 20.10.2014.

[4]Die Welt, "Die Entdeckung der Langsamkeit mit einem Traktor", 20.11.2014.

geographic South Pole on the 9 December 2015. Her adventure had been permitted by an Antarctic Treaty Party. However she was not allowed to build her snowman at the South Pole or to leave any other items in the snow. (www.tractoractor.org/). Her expedition, better known, as the Antarctica2 expedition, was followed on social media alone by more than 27 million people (www.mid-east.post.page4me/-blog/2015/06/27/) and these days the former actor is giving presentations all over Europe. As for the German explorer Martin Szwed it seems to be more difficult to decide, if he succeeded in his goals or did not. Szwed claims to have shattered the speed record for a solo ski to the South Pole in 2015. But his timeline was highly doubtful. He has revealed no GPS data, no photos can be seen on his blogs, showing him at the South Pole either. There is no proof, that he even attempted the journey. Since his return from Antarctica he is subject of two investigations by the German government. In Germany he had applied to the Federal Environmental Agency for a permit and when he was denied passage, he went anyway. The pressure to be successful has seemingly led him to cut corners, falsifying records and outright lie. (www.outsideonline.com/2047761/martin-szwed-antarctica-record-claim)

But Adventure travel does not have to be necessarily understood as hardcore pursuits. Nowadays, travellers are more aware that adventure travel encompasses a wide range of activities, including sporting events and tailor made “experiences”. One could also use the phrase “adventure curious” to describe many of today’s soft-adventure travellers. Travel agents all over the globe are planning and arranging unique experiences for their own adventure curious clients and might approach Antarctic Operators with specific ideas of their own (Kenneth Shapiro 2016).

A related problem are the new gadgets, passengers take with them on their adventures. As Unmanned Aerial Vehicles (UAV) are getting cheaper and more sofisticated, more and more Antarctic passengers would like to fly them in Antarctica to help get some aerial pictures. Experiences with UAV′s during the season 2014/2015 showed that not only there would be a need for guidelines, but also for legislation in those countries, providing gateways for Antarctic Tourism. Most operators chose to deny requests to fly UAVs or drones in Antarctica during that season, as they were not comfortable with the mitigation measures. Due partly to the work load to answer inquiries to less than understanding potential pilots by the office personnel of the tour operators and the extreme attention the launch of these devices need on board, most operators decided not to allow flights any more.

Sadly drones have been lost in Antarctica and added to the problem of garbage left behind. IAATO cautions all potential travellers to Antarctica, who are hoping to fly a drone, or Unmanned Aerial Vehicle (UAV), to check with their travel agent or tour operator before packing their device. IAATO member operators have agreed to ban the recreational use of UAVs for the 2015–2016 season in Antarctica’s unique, and often wildlife rich, coastal areas while more information is gathered about their safe and environmentally responsible use. (www.iaato.org Featured articles and events).

"In an Antarctic context, UAVs have the capability of being extremely useful to operators. For example UAVs can be used in the assessment of sea-ice conditions; crevasse reconnaissance; for outreach and education and for offering an unusual and potentially powerful perspective of the Antarctic environment. Equally, however, UAVs have the potential to cause more than a minor or transitory impact, particularly in wildlife rich coastal regions of Antarctica such as the Peninsula or Ross Sea areas. Their use could also undermine other visitor's wilderness experience if operated incorrectly or insensitively, and may generate waste if lost" (IAATO 2016)

Preliminary draft guidelines for the use of UAVs were adopted in May 2014 at the IAATO25 Meeting as a quick and efficient response to this pending management challenge. Prior to the season, a media outreach campaign was launched aimed at potential IAATO members' clients to highlight the need to contact their operator prior to bringing a UAV with them, and noting that any activity in Antarctica needed to be authorized in advance of the activity occurring. Additionally, the IAATO mandatory briefing, usually given by the Expedition Leader (EL), was updated to include reference on use of UAVs (IAATO, IP88 2015a, b).

At the same time, the IAATO Field Operations Committee had prepared new guidelines for sea kayaking and underwater activities, which included considerations about pre-requisites for engaging in the activity as well as instruction on how to operate from the ship, clothing and equipment as well as the briefing and finally the deployment in the field. These new guidelines were adopted during the annual meeting in April 2015 and incorporated into the IAATO Field Operations Manual for the 2015–2016 season and onwards. Considering, that both, underwater activities and sea kayaking had been carried out already for a good number of years, this was no surprise. At the same time, both activities were in sync with the general idea, that visits to Antarctica should also provide an educational aspect. "Underwater activities give a unique perspective of Antarctica′s marine wildlife in its natural habitat and the underwater icescape." (IP88) Whereas Sea kayaking "provides opportunities for visitors to appreciate Antarctica as a wilderness from a different perspective, allowing visitors the opportunity to enjoy the unique environment in a tranquil, quiet way." (IAATO, IP88 2015a, b).

Following the ATCM Multi-Year Strategic Work Plan: Summary of the ATCM discussions on land-based and adventure tourism, IAATO considered it helpful to offer definitions and provide a general overview of these activities as reported by their members. In their paper, IP78 presented at ATCM XXXVIII in Brazil (2014) they summarize both adventure activities and land-based tourism, within defined parameters and give an overview of any perceptible trends during the last 10 seasons. Their members are of different nationalities, and hence National Antarctic Programs do not necessarily share the same criteria for permitting these special tourism activities, they have to assess how best to assist their members in the process.

But in the past new modalities have always developed ahead of regulations, sometimes very rapidly. So it seems to be legitimate to ask: What's next in Adventure Tourism?

As an example,—although maybe a bit farfetched—we would like to mention that some tour operators are exploring the possibilities the Intersection of Food Experiences and Adventure Travel might offer. Under the title "Taste the Adventure" research reports and a database with a roundup of trends provided by an expert panel comprised of nine food tourism pioneers are already available to cater for this (Adventure Travel Trade Association ATTA 2015). According to their findings, one in four adventure itineraries revolve around food experiences, and 71 % of adventure itineraries have an experimental food focus of some kind. There are of course the traditional cooking classes and visits to wineries, breweries and distilleries, but the forum is also offering a discussion of the challenges and opportunities in combining food tourism experience with existing adventure travel products.

For some operators in the Arctic, this already means to include musk oxen and reindeer meat in their onboard menu. The closest a passenger on an Antarctic Cruise ship can get to taste the "local" food sources so far is trying reindeer from the Sub-Antarctic Island of South Georgia. The non-native reindeer have been removed from South Georgia as part of a broader Habitat Restoration Plan, and the recovered meat has been sold to the cruise ship industry and local restaurants in the Falkland Islands.[5] For passengers, who would like to follow in the footsteps of their polar heroes, this might not be enough. Especially considering how very important food has been,—both posing problems and providing pleasure to polar explorers—(Schillat 2007 and Feeney 1997), modern day tourists are quite inquisitive about recipes and tastes of the food sources in Antarctica. The publication of several books, dealing with stories of "Antarctic Cuisine" in the 1950s and 1960s, has been a success (Anthony 2012 and Cutland 2010). Both authors offer descriptions on how to prepare best penguin, cormorant and seal meat and do remind readers that it was quite common to use fresh meat on the bases in Antarctica before all species were protected.

It does not seem to be likely that any travel agent would propose to reenact a culinary experience from the 1950s, when travelling in Antarctica. But this one, like any other proposed activity would have to be approved by a National Antarctic Program in order to obtain a permit to be carried out legally. Countries might have different opinions about permitting in general, but there is also a robust framework in place at the ATS, which protects the wildlife in Antarctica in this case. At the

[5]Reindeer were introduced to South Georgia by Norwegian whalers from 1909–1925. At first the herds were controlled through regular hunts, but since the 1980s no hunting or management had been carried out and as a consequence the herds expanded substantially. Following a long period of stakeholder consultation, in 2010 the government of South Georgia & the South Sandwich Islands (GSGSSI) made the decision to remove reindeer from the island. The operation succeeded and a total of 7.685 kg of meat was recovered which has been sold to the cruise ship industry and local restaurants in the Falkland Islands. (www.gov.gs/environment/eradication-projects/eradication-projectsreindeer/).

same time, IAATO advocates and promotes operation by the membership within the parameters of the Antarctic Treaty System, including the Antarctic Treaty and the Protocol on Environmental Protection to the Antarctic Treaty, as well as IMO Conventions and similar international and national laws and agreements. It is with this mission statement, that IAATO would have to find a way to best serve and assist their members in the process of filing applications for permits and educating them about established rules.

Lately, Treaty Parties can be heard demanding that a strategic approach to tourism should be developed. They are concerned that the industry might be increasing and diversifying faster than the ATCM could regulate its activities. A strategic approach to a legitimate activity, such as tourism, should consider the question of and ultimately define the kind of tourism the Treaty Parties would want to see in Antarctica. It should focus not only on the problems raised by Antarctic tourism, but also be proactive in providing pathways for solving those problems. At the same time though, a strategic approach would need to remain broad enough to be able to provide consistency in treatment, with the ability to focus and apply that approach to specific areas of concern as they arise. Finally, any strategic approach should always be framed by the values expressed in Article 3 of the Environmental Protocol, namely the protection of the Antarctic environment for its aesthetic, wilderness and scientific values.

As priority areas the following were defined: Improving the measurement and management of cumulative impacts; Increasing the level of monitoring of tourism activities; discussions about how best to prevent or regulate the further expansion of tourism activities in Antarctica, including the development of commercial aviation and others. Many of the priority areas identified are the same areas that have been mentioned in previous working papers submitted to the ATCM, in particular the *Strategic Vision of Antarctica Tourism for the Next Decade,* submitted by the United Kingdom in 2009, and the *Report on the Intersessional Contact Group "Outstanding Questions" on Antarctic Tourism,* submitted by the Netherlands in 2012.

There are already a number of regulations in place for Antarctic tourism. However, today there is a debate if the existing regulatory framework is still fit for purpose, and parties might have to be open to discussing and considering the adoption of new instruments, legally binding or otherwise, to address the diversification of tourist activities. Possible new tools suggested included the restriction of tourism to certain activities and/or areas, limits on visitor numbers and time allowed, and the imposition of a visitor fee to go towards ensuring the protection of the environment.

And as the debate unfolds tourism is not likely to leave the centre stage of discussions in amongst Antarctic Treaty Parties, the international conservation community and the public at large.

References

Antarctic Treaty Secretariat, ATCM37, SP16. (2014). "Working Group on Tourism and Non-governmental Activities—Chairman's Proposed Agenda and Summary of Papers".

Antarctic Treaty Secretariat, ATCM38, SP15. (2015). "Working Group on Tourism and NGO Activities: Summary of Papers".

Anthony, J. C. (2012). *Hoosh. Roast Penguin, scurvy day, and other stories of antarctic cuisine* (286 p.). Lincoln, London: University of Nebraska Press.

ATTA (Adventure Travel Trade Association). (2015). (www.adventuretravel.biz/store/taste-the-adventure/)

COMNAP, ATCM38, WP22. (2015). "UAV use in Antarctica—Risks and benefits".

Cutland, G. T. (2010). *Fit for a "FID". Or how to keep a fat explorer in prime conditions* (97 p.). United Kingdom Antarctic Heritage Trust (UKAHT), [First published by: British Antarctic Survey, 1957].

Feeney, R. E. (1997). *Polar Journeys. The role of food and nutrition in early exploration* (279 p.). Washington: University of Alaska Press.

Germany and Poland, ATCM37, WP5. (2014). "UAVs and their possible environmental impacts".

Headland, R. (1994). Historical development of Antarctic tourism. *Annals of Tourism Research, 21*(2), 269–280.

IAATO, ATCM38, IP86. (2015). IAATO Guidelines for Sea Kayaking and Underwater activities.

IAATO, ATCM38, IP88. (2015). IAATO Policies on the Use of Unmanned Aerial Vehicles (UAVs) in Antarctica.

IAATO, ATCM 37, IP77. (2014). "Management of Tourism in Antarctica—an IAATO Perspective".

IAATO, ATCM 37, IP78 (2014). "Adventure Tourism: Activities Undertaken by IAATO Members", 2014.

IAATO, ATCM 36, IP98. (2013). "IAATO Guidelines for Short Overnight Stays".

IAATO, ATCM32, IP101. (2009). "Land-based Tourism Facilties", 2009.

IAATO, ATCM31, IP84. (2008). "Land-based Tourism and the Development of Land-Based Tourism Infrastructure in Antarctica: An IAATO Perspective".

Jabour, J. (2014) Strategic management and regulation of antarctic tourism. In T. Tin, D. Liggett, P. Maher, & M. Lamers (Ed.), *Antarctic futures: Human Engagement with the antarctic environment* (pp. 273–286). Springer.

Powell, R., Kellert, S., Ham, S. (2008, July). Antarctic tourists: ambassadors or consumers? In: *Polar Record*, vol. 44, issue 3, pp. 233–241.

Schillat, M. (2007). Pemmican and penguin-breast, but no pie: Daily problems of Polar explorers during the Heroic Age of Antarctic exploration. In: R. Jorge, & L. B. María(Eds.), *Antarctic Peninsula and Tierra del Fuego: 100 years of Swedish-Argentine scientific cooperation at the end of the world* (pp. 169–178). London: Taylor & Francis.

Singh, Tevjir. (2004). *New horizons in tourism: Strange experiences and stranger practices*. London: CABI Publishing. 240 p.

Shapiro, Kenneth Adventure Travel News, Kenneth Shapiro. (2016). "*Open the door: 5 Reasons why travel agents want to work with adventure companies*" ATTA (Adventure Travel Trade Association). (www.adventuretravel.biz/)

Spletstoesser, John. (2000). IAATO's stewardship of Antarctic environment: A history of tour operator's concern for a vulnerable part of the world. *International Journal of Tourism Research, 2*(1), 47–55.

Splettstoesser, J., Landau, D., & Haedland, R. (2004). Tourism in the forbidden lands: the antarctic experience. In T. Singh (Ed.), *New horizons in tourism: Strange experiences and stranger practices* (pp. 27–36). London: CABI Publishing.

The Netherlands, ATCM35. (2012). "Report on the Intersessional Contact Group "Outstanding Questions" on Antarctic Tourism".

UK, ATCM35, WP4. (2012). "The Assessment of Land-Based Activities in Antarctica".

Chapter 5
The Origins and Development of Antarctic Tourism Through Ushuaia as a Gateway Port

Marie Jensen and Marisol Vereda

Abstract From a diachronic perspective, this contribution offers an analysis of seaborne tourism to Antarctica focusing on Ushuaia as a base port and/or a port of call from 1958 to 2014. In this respect, a periodization was undertaken according to key factors that determined the tourist shipborne flows to Antarctica. The analysis starts with the first tourist voyage to Antarctica on board the vessel *Les Eclaireurs*, organized by Argentina in January 1958 and it finishes in 2014, including the 2014/2015 Austral summer season. The comparison of different sources enables a revision of the voyages, tourist flows, evolution of seasons, factors that influenced the development of Antarctic tourism and the role of Ushuaia as a gateway city. Likewise, a reference to the "tourist boom" in the 1970s is given, since it is the period when Antarctic shipborne tourism consolidates as a tourist product, being the most active decade before the sustained growth that started in the 1990s. Also, the occurrence of large cruise ships is taken into account. Finally, an integrated analysis per period is provided.

Keywords Tourist flows · Gateway · Antarctic voyages · Antarctic visitors · Government and private sectors

5.1 Introduction

Ushuaia (54°48′ S–68°19′ W), located in Tierra del Fuego, Argentina, at the southern tip of South America, is the most active gateway city to Antarctica in terms of Antarctic shipborne tourism, concentrating more than 90 % of the worldwide maritime Antarctic tourist activity.

M. Jensen (✉) · M. Vereda
Instituto de Desarrollo Económico e Innovación,
Universidad Nacional de Tierra del Fuego, Onas 450, 0910 Ushuaia, Argentina
e-mail: mjensen@untdf.edu.ar

M. Vereda
e-mail: mvereda@untdf.edu.ar

M. Schillat et al., *Tourism in Antarctica*,
SpringerBriefs in Geography, DOI 10.1007/978-3-319-39914-0_5

Antarctic tourism has shown an important increase since the beginning of the 1990s. Antarctic tourism can be defined as the commercial activity that involves the travelling of persons south of 60° S or Antarctic Treaty Area, for leisure purposes, paying special attention to motivations related to wildlife, dramatic landscapes, scientific interest and exploration history. This type of tourism requires the support given by gateways in the Southern Hemisphere in order to be able to carry out the voyages.

In this respect, Ushuaia is one of the most convenient gateways for Antarctic maritime tourism. The development of airport and port infrastructure in Ushuaia as well as a diversity of specialized services has contributed to facilitate operations of Antarctic cruise ships. This, in addition to the short distance between Ushuaia and the Antarctic Peninsula, has allowed tour operators to extend their Antarctic operating schedules.

This contribution offers a diachronic analysis of seaborne Antarctic tourist flows focusing on Ushuaia as a base port of call from 1958 to 2015. To do so, a geo-historical perspective was undertaken, enabling a periodization that starts when the first tourist voyage to Antarctica on board the vessel *Les Eclaireurs* took place in January 1958, and finishes in 2015, including the 2014/2015 Austral summer season. For the most recent periods, data were collected from the manifests[1] provided by cruise ships to the Argentine Maritime Authority (*Prefectura Naval Argentina)* while for the early stages of Antarctic tourism the book of registries at the *Prefectura Naval Argentina* (PNA 2000, 2010) was consulted. This information was compared to data taken from bibliography, to itineraries provided by tour operators and to data obtained from enquiries whenever needed.

5.2 The Role of Gateways for Antarctic Tourism

For Antarctic tourism, the role of gateway ports is fundamental. Since Antarctica is an isolated continent from a geographical point of view, it is essential to count on city ports for the development of different activities, governmental and non-governmental ones, which are carried out in this remote continent.

Particularly, Antarctic tourism needs to call at connecting ports every few days due to the necessary change over passengers and also for a diverse variety of logistic operations.

To some extent, gateways refer to the international entry and exit points to a country or region in terms of transport routes, through which tourists have to pass when entering a destination (Meyer 2004). As such, they are often associated with an international airport, major road, railway station or seaport (Pearce 2001). Among the main attributes mentioned by Burghardt (1971 in Pearce 2001),

[1]It refers to the inbound and outbound clearance documentation provided by the captain in charge of the vessel.

gateways are in command of the connections between the tributary area and the outsider world in regard to the flows of goods and people, covering long distance connections.

More specifically, the expression "Antarctic gateway" derives from the idea of "Entrance States", which refers to the countries located in the Southern Hemisphere. It originated in the Conference of Washington in 1959 when the Antarctic Treaty was signed (Guyer 2006). Bertram et al. (2007) defines an Antarctic gateway as a coastal or island port, able by its closeness to the Antarctic to benefit from and control access to Antarctic resources, including the activities that take place there (scientific support, fishing and tourism).

Due to the characteristics of Antarctic tourism, gateway cities must have suitable infrastructure for Antarctic tour operators to operate, such as an appropriate airport for passengers' arrival and departure, as well as adequate port facilities and services for vessels to prepare for an Antarctic expedition. Port services, provided by government organizations, private-sector companies or both combined, include pilotage, mooring, stowage, provisioning, stevedoring, water supply, fuel supply, garbage disposal and treatment of bilge waters and waste. Other complementary services at the Antarctic gateway are related to tourism, such as accommodations, tours, varied gastronomy, transportation to points of interest, etc. A number of facilities such as banks, medical services, shops, etc. are used by locals, tourists and tour operators alike. Additionally, human resources at the gateway city should be skilled and specialized to attend to the specific needs and requirements of Antarctic tour operators and passengers, and offer services accordingly. Finally, Antarctic gateways are also characterized by the educational, artistic and cultural expressions, which relate the gateway city to Antarctica.

Besides, another prevailing condition for Antarctic gateways is to be close in geographic proximity to Antarctica, since it enables tour operators to have a shorter period of navigation, adding at least one more voyage by the end of the season.

In regard to Antarctic tourism, five gateways are recognized: Ushuaia, Argentina; Punta Arenas, Chile; Christchurch, New Zealand; Hobart, Australia and Cape Town, South Africa. As it can be seen in Fig. 5.1, the closest gateways are Ushuaia and Punta Arenas, located approximately 1000 and 1200 km respectively from the northern tip of the Antarctic Peninsula.

5.3 The Beginning of Ushuaia and Its Relationship with Antarctica as a Gateway City

For many years, the protected waters of the Beagle Channel were frequented by sea nomads who named the bay where the city is located nowadays "*Oshovia*", which later became "Ushuaia", term that means "bay that penetrates to the West". The city stands in a sheltered bay on the northern coast of the Beagle Channel.

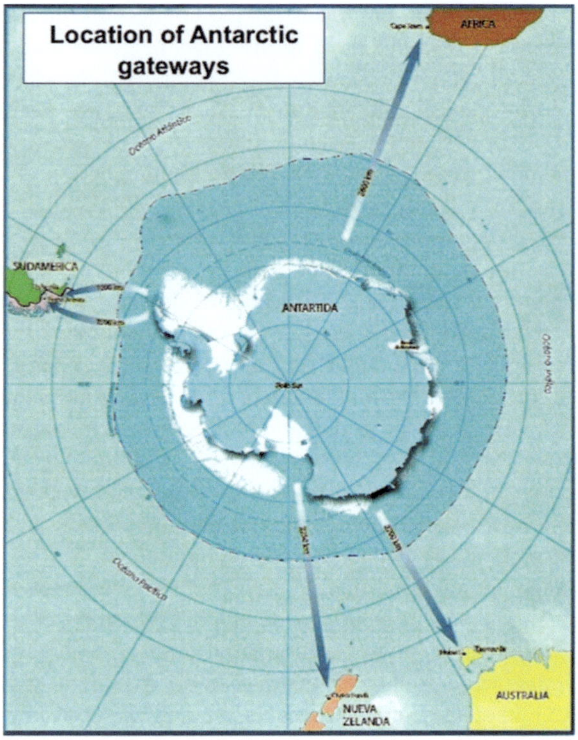

Fig. 5.1 Location of Antarctic gateways. *Source* Vereda (2004)

Even though the first known strait connecting the Pacific and the Atlantic oceans, the Strait of Magellan, was discovered by the Portuguese seaman Hernando de Magallanes in 1520, the Beagle Channel, which also communicates both oceans, was found much later, by the expedition commanded by Phillip Parker King in 1830. Fitz Roy, captain of the Beagle, confirmed the discovery communicated by Master Murray who had seen a narrow opening between Hoste and Navarino islands (south of Tierra del Fuego). In fact, he had discovered the narrows they called Murray and the channel they named Beagle after the ship.

In 1869, the Anglican mission settled down in the bay of Ushuaia, since they were looking for a place with certain characteristics: accessibility for the supply of ships even in bad weather, protection from the strong winds, etc. finding in the bay of Ushuaia the best conditions.

Between 1882 and 1884, the Argentine government developed a project whose aim was to create several navy stations along the Patagonian and Fuegian coasts, thats why a survey of the coasts of Tierra del Fuego and Isla de los Estados (Staten Island) was done in order to decide the best location for the recently created *subprefecturas* (coast-guards). Consequently, the advantages of Ushuaia as an anchorage point were noted and on 12 October 1884 the *Subprefectura* Ushuaia was created.

Although Antarctica had called the international attention due to the voyages of discovery and the economic activities of sealers and whalers, the International Geographical Congresses in London (1895) and Berlin (1899) played a major role in increasing scientific interest and public support of Antarctic research, leading to international collaboration. In this respect, Argentina was willing to cooperate and in 1902 set up the observatory located on Observatorio Island (Año Nuevo archipelago), in the South Atlantic, giving support to expeditions organized by different countries. On 22 February 1904, Argentina owned the meteorological and magnetic observatory that had belonged to the Scottish private expedition in charge of William Bruce, on Laurie Island, South Orkney (*Orcadas del Sur*) archipelago (Capdevila 2001; Genest 2001, 2005).

In terms of infrastructure, in 1903, the cadastral plan of Ushuaia already showed the governmental pier, with a depth of water of approximately 4 m (Dirección Provincial de Puertos 2007).

It was early in the twentieth century, when the first cruise ships call at Ushuaia. Several maritime companies used to sail across the Fuegian Channels aboard their huge liners. During the 1922/1923 summer season, the ocean liner *Cap Polonio* carried out voyages from Buenos Aires to Tierra del Fuego, with stops at Punta Arenas, Ushuaia, Harberton, etc. (De Agostini, n.d.). The *Cap Polonio* was a luxury liner owned by Hamburg-Sud American, the voyages were organized by the tour operator *Delfino and Co.* Three voyages of 18 days took place in the 1922–23 season, transporting a total of 1,235 passengers.[2] Voyages continued to the Beagle Channel, they rode at anchor in the bay since the pier was too small for such big ships. Other cruise ships that visited the area were the *Antonio Delfino* (1929 and 1930) and the *Monte Cervantes* (1930), the latter being the protagonist of a famous wreck in the bay of Ushuaia, and the *Monte Pascoal,* which visited the area twice (1934), closing this first cruise period in Tierra del Fuego.

With respect to the maritime links of Ushuaia and Antarctica, in 1933 the Argentine naval vessel *Pampa*, making a relief voyage at the Observatory on Laurie Island, took along a party of visitors: a journalist[3] together with his wife and daughter, a reporter and some members of the University Club of Buenos Aires; also, the American Naval Attaché and his wife and daughters were among the party as special foreign guests. Once in Ushuaia, the Attaché and his family decided not to continue the voyage towards the Antarctic since he considered that the vessel *Pampa*, with her iron hull, was not safe enough to undertake such an expedition. (Capdevila and Comerci 2013). In an article written by the journalist and published by the magazine *Caras y Caretas*, this group of people was identified as the first Antarctic visitors, and several impressions of their voyage were expressed,

[2]Data were obtained from a brochure corresponding to the 18° voyage organized by *Delfino and Co.* to the Fuegian Channels in 1929 on board the vessel *Monte Olivia*. Available at http://www.histarmar.com.ar/LineasPaxaSA/41-HamburgSud-Crucero1929MonteOlivia.htm.

[3]He worked for a very famous Argentine magazine named *Caras y Caretas*, which was published in Buenos Aires between 1898 and 1941.

demonstrating the strong sensations they experienced in Antarctica (De Souza Reilly 1933).

In 1937 the pier of Ushuaia was enlarged, reaching 750 m long (considering both sides), and a maximum depth of 8.5 m (Dirección Provincial de Puertos 2007).

In the following decades, different kinds of studies and scientific research related to Antarctica took place, increasing the existing knowledge and experience about the continent. In many occasions, special guests such as ambassadors, navy representatives, authorities, scientists and journalists were invited to join the expeditions on board Argentine naval vessels (Arguindeguy 1972).

In the 1950s, several tourist voyages to Ushuaia and the Fuegian channels took place, many of those vessels flew the Argentine flag: *Juan de Garay, Les Eclaireurs, Lapataia* and *Le Maire* (Arguindeguy 1972).

In those years, Ushuaia depended on maritime communications for the transportation of passengers as well as for the provision of supplies since regular flights started in 1948 and the main road was not opened until 1954.

As far as half of the last century, the events mentioned above demonstrate on the one hand, the tourist interest that southern and Antarctic places awakened and, on the other, the availability of knowledge, human resources and logistics that made possible the beginning of maritime tourist activities in the area.

5.4 The Beginning of Antarctic Tourism Through Ushuaia. Government Initiatives

Even though Antarctic tourism preceded the sign of Antarctic Treaty in 1959, the tourism activities were not mentioned in it at all. Nevertheless, later on, measures, decisions and resolutions related to tourism were adopted at the Antarctic Treaty Consultative Meetings (ATCM).

Maritime tourism towards Antarctica started in 1958, on the initiative of Argentina when organized two journeys on board the vessel *Les Eclairerus* (Reich 1980; Pierrou 1981; Headland 1994). By the end of the year 1957, a call for applications was published by the Naval Transport Command (*Comando de Transportes Navales*) in order to analyse the requests of people interested in visiting Antarctica according to their nationality, occupation and purpose of visit. This idea was taken in order to set up a diverse group of travellers (Nichols 1959). Therefore, the first voyage to Antarctica departed from Ushuaia on 16 January 1958, carrying 98 passengers, they visited the following Naval Detachments: Decepción, Cámara, Brown and Melchior. These first Antarctic visitors unveiled a commemorative plaque in the main building of Decepción, since it was the first place in Antarctica where they landed.[4] Also, a travel log was kept during the voyage, called "The

[4]nformation was taken from the newspaper *La Nación*, March 16th 1958, Sect. 2, p. 3, Buenos Aires.

Penguin Tourist" that enabled visitors to keep up with the news on board. In fact, this turned out to be the first journal published in Antarctica (Jutronich 1958).

The second voyage departed from Ushuaia, too, on 31 January, with 96 passengers on board. It returned on 11 February. From Buenos Aires to Ushuaia people were transferred in a DC-4 plane, belonging to the Air Force (Dirección de Turismo Territorial 1986). In this respect, national authorities were convinced that the government had to demonstrate that it was possible to develop tourism to Antarctica (Pierrou 1981), based on the experience already gained through exploration, surveying, researching, etc.

Arguindeguy (1972), while describing the tasks undertaken by the 23rd Argentine Antarctic campaign, pointed out the support given by the Naval Antarctic Group to the '*Les Eclaireurs*', in her first voyage with tourists to Antarctica.

On 23 January 1959, the vessel *M/B Yapeyú*, belonging to the Argentine Overseas Fleet (*Flota Argentina de Navegación de Ultramar*) made a 16-day voyage, carrying 262 passengers, who departed from Buenos Aires to the South Shetland Islands (*Islas Shetland del Sur*) and the Antarctic Peninsula, making a stop in Ushuaia on the way back, on 6 February. The *Yapeyú* was a passenger vessel that had undertaken a round-the-world voyage in 1956. Pierrou (1981, p. 740) remarks that during that season among the assigned missions to the Antarctic Naval Force, No. 9 specifies the intention to "(…) give full support to the planned tourist voyages".

In sum, in the last years of the 1950s, three tourist voyages took place, carrying 456 passengers. They account for the 0.08 % of the total voyages and the 0.09 % of the total visitors (considering the whole period up to the 2014/2015 season), in regard to Antarctic seaborne tourism. These voyages had been the only ones organized before the Antarctic Treaty was signed, and were undertaken by the government, operating under the flag of the organizing country.[5] These first experiences could demonstrate both the technical viability to make the voyages and the interest of tourists to visit remote areas. Furthermore, the exploration feeling as well as the beauty encountered highly satisfied visitors' expectations.

5.5 Private and Government Sectors Initiatives to Follow up Maritime Antarctic Tourism in the 1960s

Tourist voyages started in the 1965/1966 season. In January 1966, *Lindblad Travel*, registered in New York, together with *Astra* travel agency, located in Buenos Aires, chartered the vessel *Lapataia* from the Naval Transport Command (*Comando de Transportes Navales*) in order to organize the "First International Survey Voyage to Argentine Antarctic Sector—January 1966" (Borten 1974, p. 4). The ship left

[5]A similar voyage was organized by Chile in 1959, aboard the naval vessel *Navarino*, carrying 84 passengers (Reich 1980; Palazzi 1993). This journey did not register a call at Ushuaia.

Buenos Aires on 13 January, calling at the ports of Mar del Plata and Ushuaia and, once in Antarctica, passengers visited some research stations (Brown, Decepción and Esperanza) (Palazzi 1993). During this voyage, *Lapataia* was accompanied by another Argentine naval vessel, the *Irigoyen* (Palazzi 2005). When they were back, the travel agency *Astra*, offered a cocktail to the visitors, the senior crew and special guests; the Argentine well-known newspaper *La Nación*[6] recorded the event under the headline: "Warm impressions of the first Antarctic tourists". Some passengers' comments can be pointed out

- "I enjoyed every minute. I have travelled round the world several times and I have never seen anything like this. Icebergs were more impressive than the Rhin castles, and we saw lots of whales". (Mrs Sweeny).
- "Even though it was hard to gather the U$S 3,000 for the voyage, once in Antarctica I wished the ship would have been trapped in the ice. It is unbelievable to fall in love with glaciers, but it happens". (Miss Flather, an astronomer from Wisconsin University).
- "This continent has lots of tourist possibilities… a lot of people have already travelled around the world and this is a new place to visit". (Mr. Jaime McMahon, manager of an aviation company).

Borten (1974, p. 23) assures "Following the spectacular interest raise in the Argentine and international press publishing this human quest for discovery and adventure, the Lindblad tourist expedition projected on to the world's screen Antarctica's magnificent scenery, unusual wildlife and scientific findings".

In regard to the participation of the private sector in the organization of tourist voyages in Antarctica and being aware of the potential growth of the activity, the IV Antarctic Treaty Consultative Meeting that took place in Santiago, Chile, in 1966, issued the first Recommendation related to tourism (ATCM IV 1966).

In 1967 *Lindblad Travel* rented again the vessel *Lapataia* and made two voyages to Antarctica in January and February, these voyages were called: "Argentine Antarctic Sector and Tierra del Fuego in the International Tourism Year—1967". The first departed from Buenos Aires to Antarctica on 8 January, calling at the port of Ushuaia whereas the second left Ushuaia to Antarctica on 3 February, calling at Ushuaia on the way back to Buenos Aires (Borten 1974).

During the following two seasons *Lindblad Travel* chartered Chilean and Danish vessels, trying out new itineraries. These voyages did not register calls at the port of Ushuaia.

For the 1968/1969 season, the National Tourism Board (*Dirección Nacional de Turismo*) jointly with the Antarctic National Board (*Dirección Nacional del Antártico*) and the collaboration of the Navy, organized a series of four voyages named "White Continent Cruises". Voyages were made with the vessel *Libertad*,[7]

[6]Information taken from the newspaper *La Nación* December 13th, 1966, p. 9. Buenos Aires.

[7]In the beginning she was both a reefer and passenger vessel, but later she was modified to carry only passengers, with a capacity for 400.

belonging to the Argentine Maritime Line Company (*Empresa Líneas Marítimas del Estado*). Of a total of 1,139 passengers, most were Argentine or foreigners living in Argentina, approximately 70 were Europeans, some from North America and South America (Dirección Nacional del Antártico 1971).

In an article called "Antarctica: tourist cruises" (Dirección Nacional del Antártico 1971) there is an interesting paragraph that points out how the Antarctic tourist experience was changing in a positive way, basically due to modern transportation and more security.

In brief, during this decade, eight voyages took place carrying 1,394 passengers that represented 0.22 % of total voyages and 0.27 % of total passengers (considering the whole period up to the 2014/2015 season). The main characteristics of this decade refer to the continuity of voyages encouraged by the government on the one hand and, on the other, the beginning of the private sector organized trips, led by Lars Eric Lindblad who chartered vessels from different countries. He experienced new itineraries that departed from diverse places: Buenos Aires and Ushuaia (Argentina), Punta Arenas (Chile) and Lyttelton (New Zealand). Vessels were registered under the flags of their respective countries.

5.6 The First Antarctic Tourist "Boom" in the 1970s

During this decade, the tourist activity was distributed between the private-sector initiatives and those promoted by different government agencies from Argentina.

The National Tourism Board was the government entity that encouraged tourism to Antarctica and, in the 1969/1970 season, together with the Maritime Government Line Company and travel agencies (Dirección Nacional del Antártico 1971), organized only one voyage of 16 days aboard the passenger vessel *Río Tunuyán*.

In the following season (1970/1971), two voyages were made aboard the same vessel, 90 % of the passengers were Argentine or foreigners who lived in Argentina while a 10 % were Europeans and North and South America residents (Dirección Nacional del Antártico 1971).

In the 1971/1972 season, the National Tourism Board organized two voyages aboard the vessel *Libertad* (that belonged to the Maritime Government Line Company). The first voyage departed 18 December 1971 with emphasis on the "Midnight Sun" and on "a real white Christmas". Besides, the National Tourism Board stressed the fact that they were trying to intensify tourism in Antarctica since it was very difficult for foreign companies to do so due to distance and air-ticket cost factors, which were very expensive to cover (Dirección Nacional del Antártico 1971).

During the 1972/1973 season, the National Tourism Board organized two voyages again, that time aboard the vessel *Libertad*. These voyages were the last ones made by this vessel since it was put out of action by the end of 1973.

The National Under Secretary of Sports and Tourism (*Subsecretaría de Deportes y Turismo de la Nación*), chartered the Panamanian flagged vessel *Regina Prima*, which belonged to the Greek company *Chandris Cruises*, to carry out six voyages in the 1974/1975 season. Only the first one departed from Buenos Aires and ended in Ushuaia while the others operated from Ushuaia as the base port (Burbridge 1984). Passengers travelled to Río Grande (north of Tierra del Fuego) by plane and then they were transferred by bus to Ushuaia. Each voyage carried about 520 passengers.

In the 1975–1976 season, the National Under Secretary of Sports and Tourism chartered again the vessel *Regina Prima* and made seven voyages. These were the last voyages aboard large vessels organized by Argentine national government agencies.

The private-sector involvement in the organization of voyages started a new period in Antarctic seaborne tourism, as Borten states

> (…)1970 marked a new date in Antarctic tourism when Lindblad, following experiences gained on earlier trips, offered to this conservation-minded clientele, a specially designed expedition ship, the MS 'LINDBLAD EXPLORER' built in Finland. Constructed to cope with ice packs and constricted waterways of the Antarctic, the ship also boasts a double hull and carries inflatable rubber boats, the famous Zodiacs to land passengers on shore. It represents a new travel concept of selected "cruising expeditions" and provides the ultimate comfort and security for 90 passengers. (Borten 1974, pp. 23–24).

In a sense, the "expedition cruise" term was consolidating in the industry as a new concept. It refers to the organization of voyages to remote destinations where the ship provides not only the access to the place but also the stay, covering all the passengers' needs. The ship manager organizes the land-based activities, including lectures on board in charge of specialists on different topics.

In the 1971/1972 season, *Lindblad Explorer* made two voyages, the first one lasted 31 days, it departed from Cape Town to visit the Antarctic, then it called at the port of Ushuaia. The second voyage left from Ushuaia. During this second trip she stranded in the South Shetland Islands (*Islas Shetland del Sur*) and passengers and crew were rescued by the Chilean naval ship *Piloto Pardo* (Borten 1974). The vessel was assisted by the ships *Yelcho,* belonging to Chile, and *Zapiola* and *Bahía Aguirre*, both Argentine (Coli 2003). "The rescue was not only one of the finest rescues at sea ever, according to seasoned experts, but also a rare and memorable example in human relations" (Borten 1974, p. 24).

In the 1972/1973 summer season, the vessel *Lindblad Explorer* made five voyages, the first departed from Mar del Plata (Argentina) and after visiting Antarctica she called at Ushuaia from where the other voyages operated.

During the 1973/1974 season, the *Lindblad Explorer* departed from the port of Bluff (New Zealand) to Antarctica, ending in Ushuaia. This voyage lasted 35 days and it became the first "Antarctic circumnavigation". Then, Ushuaia operated as a base port for the other two voyages, which visited not only the Antarctic Peninsula, but also other surrounding areas.

The *Lindblad Explorer* continued with Antarctic journeys, making between two and four voyages during the following seasons in this decade.

In the 1972/1973 season the Spanish company Ybarra organized a voyage aboard the vessel *Cabo San Roque*, carrying 720 passengers, divided into two classes on the ship.

In the 1973/1974 season, the same company made two voyages throughout the season one aboard the ship *Cabo San Roque* and the other aboard the *Cabo San Vicente*, both called at the port of Ushuaia. In the following season the *Cabo San Roque* added another voyage.

In January 1973, the vessel *Enrico C*, belonging to the *Costa Line Fleet*, called at the port of Ushuaia carrying 888 passengers, coming from Antarctica. The following year, the *Enrico C* returned to Antarctica and also called at Ushuaia, with 803 passengers on board (PNA 2000; Palazzi 1993). In this respect, the *Costa Line Fleet* published on its web site: "The 1960s was a very positive decade, and to the traditional itineraries around South America and the Caribbean, cruises to the Mediterranean, Black Sea, Brazil, Uruguay and Argentina down to the Straits of Magellan and Antarctica were added".[8]

Another cruise company, the *Society Expedition Cruises*, started operating in 1977 with the vessel *World Discoverer*, making two voyages in the 1977/1978 season, carrying 282 passengers (Palazzi 1993).

In this decade, 63 voyages took place, carrying 16,824 passengers, which represented 1.74 % of the total of voyages and the 3.24 % of the total of passengers (considering the whole period up to the 2014/2015 season).

This tourist "boom" in the 1970s is characterized by the active participation of the government as well as the private-sector involvement. Voyages organized aboard large ships played a significant role in this decade. A specially designed vessel to navigate in polar waters initiated her operations in Antarctica. Besides, some vessels operated under convenience flags.

5.7 The Impasse of the 1980s

During this decade, Antarctic seaborne tourism decreased considerably. In the 1979/1980 season, the National Tourism Board organized their last voyage aboard the *Bahía Buen Suceso*. From this season onwards, the private sector took the initiative and, therefore, in 1980/1981, the travel agency *Antartur,*[9] domiciled in Ushuaia, organized a voyage aboard the vessel *Bahía Buen Suceso,* carrying 70 passengers (Burbridge 1984). Two other voyages took place during the following seasons, carrying 139 visitors (Dirección de Turismo Territorial 1986). In this

[8]Costa Cruise Fleet. The history of the company. Available at: http://www.costacruisesasia.com/B2C/SharedResources/Corporate/History/pdf/STORIA_COMPLETA_E.pdf (Para 8).

[9]This company was founded by Gustavo Giró Tapper, an Antarctic expert who was in charge of two Antarctic research stations (San Martín and Esperanza). Besides, he participated as the second chief in command of the first Argentine terrestrial expedition to the South Pole, named "Operation 90" in 1966 (*Operación 90*).

opportunity, several Ushuaia dwellers could visit Antarctica, like old settlers and flag bearers from local schools (E. Giró, personal communication, September 15, 2007).

Later, the voyages were cancelled since the economic benefit was not the one expected and besides, the vessel was sunk during the South Atlantic conflict (1982). For the following two seasons there was not any Antarctic tourist vessel calling at the port of Ushuaia (Burbridge 1984).

Antartur restarted voyages to Antarctica in the 1985/1986 austral summer season. This time they chartered partially berths from the polar vessel *Bahía Paraíso*[10] (owned by the National Antarctic Direction) in order to carry out a voyage with particular characteristics. The voyage departed from Ushuaia on 7 February 1986, and some special guests went aboard, who were related to the participants of the Swedish expedition that had taken place in 1901, such as relatives of Nordenskjöld, Larsen and Sobral. They paid tribute to the members of the expedition at Snow Hill Island (*Isla Cerro Nevado*; E. Giró, personal communication, 15 September, 2007).

In the 1986–1987 season two voyages that lasted 11 days were made, carrying 155 passengers. Several activities were organized: landings, helicopter over flights, excursions on polar tracked vehicles and dog-sledding tours.[11]

During the following summer season four voyages were organized, taking along 379 passengers (Rodríguez 1997) and other four voyages took place in the 1988/1989 season. Sometimes, small groups of tourists stayed for a couple of days at Esperanza Station performing different activities, it was also possible to rent cross-country gear there (E. Giró, personal communication, 15 September, 2007). Most of the staff was recruited in Ushuaia.

These voyages were cancelled after the *Bahía Paraíso* ran aground in the Bismarck Strait on 28 January 1989, sinking days later. The Argentine polar vessel *Almirante Irizar*, the oceanographic Spanish ship *Las Palmas* and a tugboat, rescued passengers and crew (Palazzi 1993).

Antartur developed a handbook named “Antartur Antarctic Visitors' Handbook”, where rules of conduct were specified according to legal standards of Argentina as an Antarctic Treaty signatory country. Several guidance recommendations in regard to behaviour were clearly stated, preceding the Recommendation ATCM XVIII-1 adopted in Kyoto in 1994, about guidance for visitors and for those organizing and conducting tourism and non-governmental activities in the Antarctic.

The vessel *Lindblad Explorer* made three voyages during the 1980/1981 and two in the following summer season calling at the port of Ushuaia. During the two seasons after the South Atlantic Conflict (1982) no movements of Antarctic voyages through Ushuaia were registered (Burbridge 1984).

---

[10]The polar vessel *Bahía Paraíso* had 80 berths for tourists, a dining room, a living room and bar and a hospital with a surgery room.

[11]Antartur: Antarctica. Programme and itinerary (*Antártida. Programa e itinerario*). 6th and 7th voyage departing on December 21st 1986 and on January 3rd 1987.

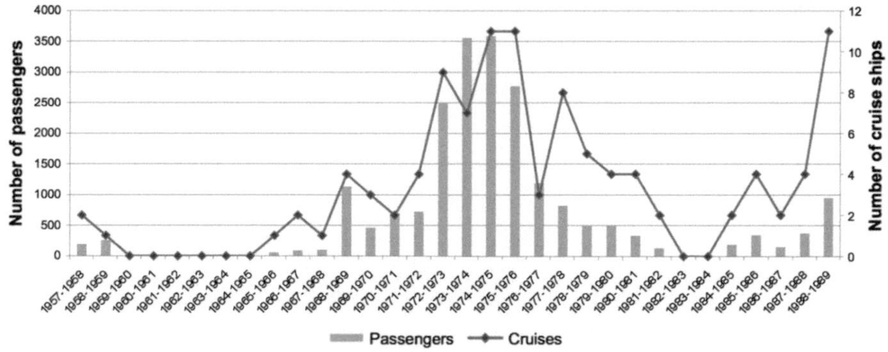

Fig. 5.2 Vessels and passengers movement through Ushuaia. 1957/1958–1988/1989 summer seasons. *Source* Authors' own, based on Jensen and Daverio (2008)

The vessel *Society Explorer*[12] organized two voyages for the 1985/1986 summer season.

In the 1988/1989 season the Russian flagged vessel *Antonina Nezhdanova* started journeys to Antarctica through the port of Ushuaia, making seven voyages. With these voyages the sustained increase of Antarctic seaborne tourism began during the following decade.

In sum, during the 1980s a total of 33 voyages took place, carrying 3,017 passengers that represented the 0.91 % of the total of voyages and the 0.58 % of the total of passengers (considering the whole period up to the 2014/2015 season).

Both the government and the private-sector activities organized from Argentina finished by the end of this decade. Also the calls made by vessels at the port of Ushuaia were suspended in the seasons that followed the South Atlantic Conflict. Tour operators experienced new itineraries operating from different ports located in South America, Australia and New Zealand. Also, some vessels operated under convenience flags.

In Fig. 5.2 the evolution of both vessels and passenger movements can be seen. It summarizes the final numbers of the first 30 years of Antarctic seaborne tourism through the port of Ushuaia. It shows the variations in the first years, the 1970s boom and the 1980s decrease. The 1977/1978 was the longest summer season in terms of voyages, lasting 82 days (see Fig. 5.3). Three decades were needed for the development of seaborne Antarctic tourism in order to be able to consolidate in the following seasons.

[12]When the ship was built, she was named *Lindblad Explorer* in 1969. Later, renamed as *Society Explorer* in 1982 when she passed to Vienna International Shipping Corporation and finally, as *Explorer* in 2004 when owned by GAP Adventurers.

5.8 The Sustained Increase of Antarctic Seaborne Tourism and the Consolidation of Ushuaia as a Gateway in the 1990s

In the 1990s, Antarctic seaborne tourism was organized by initiatives coming from the foreign private sector. Several exogenous and endogenous factors highly contributed to this development. Among the first ones, the availability of vessels to be chartered in the market as well as skilled human resources in polar issues can be mentioned; in regard to the second ones, suitable infrastructure developed in Ushuaia and a more specialized organization from the perspective of government institutions as well as business environment, helped the consolidation of this activity.

In the first season of this decade eight voyages, carrying 960 passengers, were made. During the following season, the vessel *Illiria* organized seven voyages, also the *World Discoverer* and the *Society Explorer* contributed with some journeys. The large-sized vessel *Ocean Princess* made two voyages (Palazzi 1993).

In the 1991/1992 season, voyages were organized aboard the vessels *Illiria*, *Society Explorer*, *World Discoverer*, *Akademik Boris Petrov*, *Vistamar*, *Professor Molchanov*, *Ocean Princess* and *Daphne*, showing a significant increase in the number of vessels operating through Ushuaia. The rise in numbers was also very important regarding voyages and transported passengers. The season lasted 105 days and it continued growing until the 1996/1997 summer season, reaching 147 days (see Fig. 5.3).

It is important to point out the place that the former USSR research vessels occupied once they became available to be chartered after the fall of the Soviet regime. In this decade, they accounted for most of the seaborne Antarctic voyages with 17 vessels operating throughout the season.

The Russian flagged vessel *Akademik Ioffe* was the first one performing 12 voyages within a tourist season (1993/1994). In the same period, the *Marco Polo* started operating (she continued until 2009) carrying up to 600 passengers and making between two or four voyages per season.

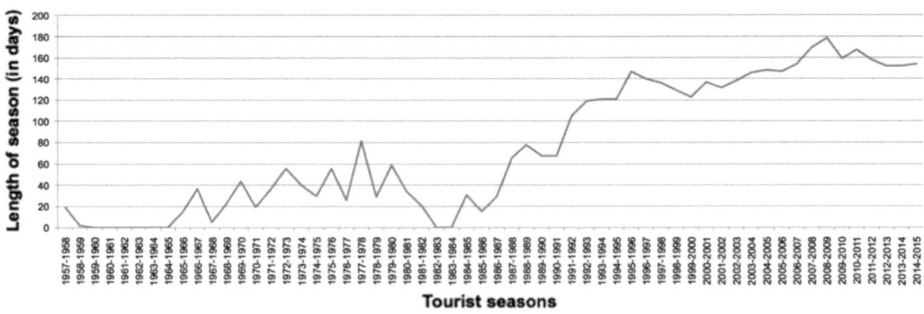

Fig. 5.3 Extension of tourist seasons from the 1957/1958 to 2013/2014 austral summer seasons. *Source* Authors' own, based on Jensen and Daverio (2008), Vereda and Jensen (2014), Argentina (2015)

In this decade, Ushuaia turned into the most active gateway to Antarctica-bound tourism, concentrating the 90.45 % of the world traffic in the 1994/1995 season. In this respect, there are some factors to be considered that contributed to this consolidation

- The Antarctic Peninsula was the preferred area for tour operators due to the presence of wildlife, moderate climate, location of research stations and the proximity to South American cities (National Research Council 1993).
- The closeness of Ushuaia to the Antarctic Peninsula allowed tour operators to add an extra voyage by the end of the tourist season.
- The building of the international airport in 1995, which was designed to allow the operation of large aircrafts as well as the construction of the air terminal (1997), favoured the accessibility and intermodal transport.
- The expansion of the commercial pier in 1999 allowed an increase in the number of berths.
- In 1992 the Antarctic Information Board (*Oficina Antártica*), a government unit under the provincial agency of tourism (*In.Fue.Tur*) was created to provide tourist information and to facilitate the operations of Antarctic tour operators.
- The specialization of port and tourist services also contributed to give more specific support to tour operators.

Besides, external factors also contributed to the sustained increase of tourist flows to Antarctica

- Broad diffusion of Antarctic tourism.
- New motivations regarding nature and a quest of new and remote places.
- Increase in the number of tour operators specialized in Antarctica as well as of cruise liners.
- Availability of suitable vessels to navigate polar waters, particularly the former USSR research vessels.

Together with the increase of voyages and passengers activities diversified and tourism became a sensitive topic of discussion at the ATCMs.

Also in this decade, the International Association of Antarctic Tour Operators (IAATO) was formed, most of the tour operators that organized voyages to Antarctica through Ushuaia belonged to this association.

Since search and rescue operations in Antarctica require special precautions, from the 1998/1999 summer season on, Argentina and Chile agreed on participating jointly and alternatively in their Antarctic areas of responsibility, giving support to all who operate in the Antarctic Peninsula. This special group is known as the "Combined Antarctic Naval Patrol".

In regard to this decade we can conclude that 585 voyages were made, carrying 55,013 passengers, which represented the 16.16 % of the total of voyages and the 10.61 % of the total visitors (considering the whole period up to the 2014/2015 season). In this decade a significant increase in the number of vessels, voyages and passengers is produced since the foreign private-sector initiative became involved. Besides, most of the vessels that operated in this decade carried between 50 and 100 passengers and an important number were Russian flagged.

5.9 The Development of Antarctic Seaborne Tourism by the End of the Millennium and the Beginning of the Twenty-First Century

During the 1999/2000 summer season several special itineraries were designed in view of celebrating the End of the Millennium in a unique remote destination, associated to the idea of the "end of the world". This motivation produced an important increase of the number of voyages, but especially of the number of passengers, reaching a 50 % more than the previous season. This increase in the number of passengers was also due to the operation of large vessels, such as *Rotterdam VI, Aegean I, Ocean Explorer* and *Marco Polo.*[13] This growth continued the following seasons, with more tour operators, vessels, voyages and passengers.

The tour operator *Antarpply Expeditions*, domiciled in Ushuaia, purchased a vessel, which was set up to carry passengers, with a capacity for 88 passengers and prepared to navigate Antarctic waters. The vessel was named *Ushuaia* and she made the first voyage in the 2002/2003 season. From the following season onwards she carried out between 9 and 14 voyages, completing 159 voyages by the end of the 2014/2015 season. Most of the crew is Argentine and Chilean and the staff from Argentina, many of them living in Ushuaia. This way, a local company started Antarctic voyages.

From the 2005/2006 to the 2011/2012 summer season the vessel *Antarctic Dream* (former *Piloto Pardo*), that belonged to the Chilean tour operator *Antarctic Shipping*, operated regularly. She made 88 voyages. *Antarpply Expeditions* and *Antarctic Shipping* are the only regional companies that operated through the port of Ushuaia in this century.

The presence of Russian flagged vessels is still significant, since during the 2008/2009 9 vessels still flew that flag.

More large vessels started to include Antarctica in their itineraries. Voyages either started in South American ports or made world cruises, among these vessels the following can be mentioned: *Ryndam, Crystal Symphony, Amsterdam, Royal Princess, The World, Insignia, Saga Rose, Discovery, Regal Princess, Golden Princess, Artemis, Prinsendam, The Topaz, Azamara Journey, Star Princess, Insignia, Seven Seas Mariner, Veendam, The Oceanic, Infinity* and *Zaandam.*

The first voyage of a Very Large Cruise Vessel class (VLCV), the *Golden Princess* took place in the 2006/2007 summer season. This vessel of 109,000 GT and 2,636

---

[13]Non-IAATO vessels: Aegean I, Ocean Explorer, Marco Polo and Rotterdam VI (IAATO 2000).

lower berths, made a voyage along the coasts of South America and Antarctica. The *Golden Princess* and the *Star Princess*, both belonging to the *Princess Cruise Company*, are the biggest cruise ships ever navigating Antarctic waters.

The 2007/2008 was the busiest season, with 44,605 passengers travelling to Antarctica by boat. The following season, 2008/2009 was the longest, reaching 179 days (see Fig. 5.3). This expansion can be explained due to the operation of the Russian flagged icebreaker *Kapitan Khlebnicov* that started the voyages very early in the season, in October, in order to visit an emperor penguin colony in the Weddell Sea.

In one of the voyages (November 2009), bad weather caused the sea-ice to compress around the ship making it difficult for the vessel to manoeuvre effectively, after a few days the icebreaker was able to navigate the pack of ice and get back to Ushuaia with six days of delay (IAATO 2009). This icebreaker stopped operating in Antarctica from Ushuaia by the end of 2010.

Concern to preserve Antarctica as well as to achieve a sustainable tourism development has been in the ATCMs agenda[14] for many years. IAATO participates as an invited expert in annual ATCMs since 1992. In general, vessels carrying more than 500 passengers did not make any landings in Antarctica according to IAATO's regulations.[15] The limitations in the number of passengers carried by a vessel to make landings were discussed at the ATCM that took place in 2007, where a Resolution concerning "Ship-based Tourism in the Antarctic Treaty Area" was adopted. Finally, in 2009 the ATCM adopted Measure 15 stating that Parties should require their operators organizing tourist or other non-governmental activities in the Antarctic Treaty area to refrain from making any landings in Antarctica from vessels carrying more than 500 passengers and in the case of vessels carrying 500 or fewer passengers to coordinate with each other with the objective that not more than one tourist vessel is at a landing site at any one time; to restrict the number of passengers on shore at any one time to 100 or fewer, and to maintain a 1:20 guide-to-passenger ratio. (ATCM XXXII 2009).

Security is also a topic of discussion in response to the perceived threats to ships and port facilities in the wake of the September 2001 attacks in the United States. Through an amendment to the SOLAS[16] Convention, the International Ship and

---

[14]From 2004 onwards, Antarctic Treaty Consultative Parties agreed on forming a working group to discuss tourism matters. Two Antarctic Treaty Meetings of Experts took place. The Madrid Protocol establishes environmental management in Antarctica. Besides, the ATCM has agreed on several measures related to tourism.

[15]The following paragraph was noted in the IAATO's information paper *Overview of Antarctic Tourism Activities* (IAATO 1999, p. 6): "The significance of the size of the vessel as it relates to potential impact, contingency planning and emergency response is complex, and factors such as choice of landing site, number of landings, type of fuel carried may be of equal or greater importance than the size of the vessel. These are not issues that are easily resolved. In particular, IAATO members are concerned that potential cumulative environmental impacts and emergency response requirements of very large vessels could adversely impact on the principles of safe and environmentally responsible travel to Antarctica".

[16]Safety of Life at the Sea (1974).

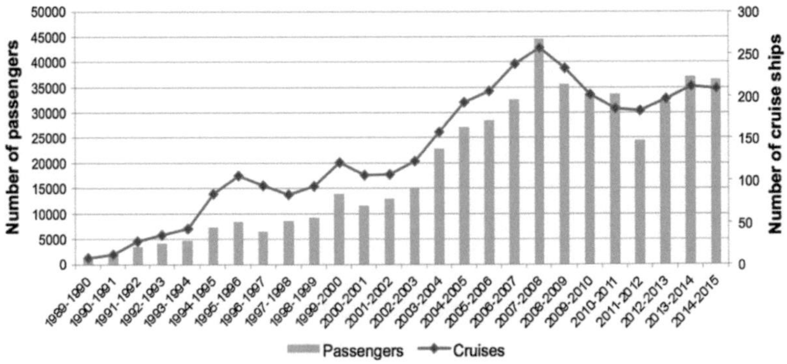

Fig. 5.4 Movement of vessels and flows of passengers through Ushuaia. 1989/1990 to 2014–2015 austral summer seasons. *Source* Authors' own, based on Jensen and Daverio (2008), Vereda and Jensen (2014), Argentina (2015)

Port Facility Security Code (ISPS) was adopted as a comprehensive set of measures to enhance the security of ships and port facilities, coming into force in 2004. The port of Ushuaia got the ISPS code in the same year.

Since the increase in the number of voyages and vessels operating in the Antarctic, incidents and accidents became more frequent. The sinking of the vessel *Explorer*[17] in November 2007 was the most remarkable accident in this century. After that, new measures were adopted with regard to the environment and the safety of life at the sea. One of the adopted measures was the amendment to Annex 1, MARPOL (IMO) that banned heavy fuel South of the 60° S and entered in force in August 2011, affecting large vessels which showed a decrease compared to previous summers seasons. Another way to diminish risks is related to widen knowledge on Antarctic navigation.

In brief, during the first decade of the twenty-first century 1,741 voyages were made, carrying 244,582 passengers, which account for 48.08 % over the total voyages and for 47.16 % over the total visitors (considering the whole period up to the 2014/2015 season).

In Fig. 5.4 the evolution of vessels and tourist flows in the last 25 years can be seen, showing a significant increase from the 1990s onwards. The celebration of the end of the millennium produced an increase during the 2000/2001 season. In this decade, the highest increase happened in the 2007/2008 season, while an important decrease took place afterwards, which may have been influenced by the international financial crisis as well as by IMO regulations that entered into force in 2011.

Throughout this period, Ushuaia maintained its role as the most active gateway city.

[17]The *Explorer* was the former *Lindblad Explorer/Society Explorer.*

5.10 Current Characteristics of Antarctic Seaborne Tourism Through Ushuaia

Seaborne tourism has had a more significant growth compared to other forms of tourism in Antarctica. According to the way of operation, different types of seaborne tourism can be identified

1. Embarkation and disembarkation in Ushuaia (both operations take place at the port of Ushuaia).
2. Embarkation and/or disembarkation in Ushuaia (vessels either departed from or arrive in Ushuaia).
   a. Embarkation and/or disembarkation of Antarctic visitors that either embark or disembark at other ports.
   b. Embarkation and/or disembarkation of Antarctic visitors that either embark or disembark in Antarctica in connection with an air service.
3. In transit (vessels that travelled to the Antarctic Peninsula as part of a longer voyage in South America or a more extensive itinerary).

In the first case, the operations in Ushuaia as a base port are important for the city since tour operators require a series of services to be provided to the vessel, the crew and passengers. In this respect, 17 vessels ran regular voyages to and from Antarctica operating throughout the 2014/2015 season from Ushuaia, making between 7 and 14 voyages. The vessel *Ushuaia* reached the maximum of voyages, completing 14 during several seasons.

Figure 5.5 shows a breakdown according to ships' modes of operation in Ushuaia. There is a slight recovery of operations that took place in Ushuaia after the international financial crisis (2008/2009), and an increase in the number of visitors that only embarked or disembarked in Ushuaia once due to the participation of other gateways and an increase in the positioning voyages that include Antarctica while approximating the destination can be seen. Besides, a significant growth in the

Season	Embarkation and disembarkation in Ushuaia	Embarkation or disembarkation in Ushuaia		In transit in Ushuaia	Total
		Seaborne	Air-cruise		
2009-2010	19,084	2,641	96	11,705	33,526
2010-2011	18,030	1,065	114	14,447	33,656
2011-2012	19,504	1,146	119	3,656	24,425
2012-2013	21,046	2,933	646	7,369	31,994
2013-2014	22,153	2,882	648	11,481	37,164
2014-2015	22,321	2,413	611	11,280	36,625

Fig. 5.5 Breakdown according to ships' modes of operation in Ushuaia, 2009/2010–2014/2015 Austral summer season. *Source* Authors' own, based on Vereda and Jensen (2014), Argentina (2015)

air-cruise mode from Punta Arenas, Chile to Presidente Frei Station in the South Shetland Islands, where the Antarctic itinerary either starts or finishes by boat took place. In this case, vessels operate in the Antarctic Peninsula region until they need to be provisioned, returning to Ushuaia to do so. With regard to vessels in transit in Ushuaia a remarkable decrease can be noted during the season after the IMO regulation had come into force, some summer seasons later numbers could recover.

Most tour operators that commercialize Antarctica either own their vessels or charter them from different ship owners. Only two tour operators are located in the region, one is domiciled in Argentina (*Antarpply Expeditions*) and the other one in Chile (*Antarctica XXI*, air-cruise mode of operation), all the other companies are based in North America, Europe and Oceania. Most tour operators specialize in Polar Regions including sometimes other remote destinations and they are IAATO's members.

For the 2014/2015 season the following tour operators integrated the association membership directory: *Abercrombie & Kent, Antarctica XXI, Antarpply Expeditions, Aurora Expeditions, Cheeseman's Ecology Safaris, G Adventures, Oceanwide Expeditions, One Ocean Expeditions, Polar Latitudes, Quark Expeditions, Travel Dynamics International, Waterproof Expeditions* and *Zegrahm Expeditions*. In recent years, the fleets of vessels started changing, decreasing the number of small ships (fewer than 100 passengers) in favour of the increase of bigger ships (between 100 a 200 passenger carrying capacity). Besides, the number of Russian flagged research vessels also decreased, for the 2013/2014 season only the *Akademik Ioffe* and *Akademik Sergey Vavilov* operated. Nevertheless, the *Ortelius*,[18] Polar *Pioneer*[19] and *Sea Adventurer*,[20] which also belonged to the former USSR, changed their registries and names and continued operating during the mentioned season.

There is a group of vessels formed by cruise lines which are also members of the Cruise Lines International Association (CLIA), the world's largest cruise industry trade association: *Celebrity Cruise Line, Compagnie du Ponant, Hapag Lloyd Cruises, Holland America Line, Lindblad Expeditions, Hurtigruten, Azamara Club Cruises* and *Silversea Cruises*. Some of them operate as expedition cruises and others only navigate Antarctic waters according to diverse itineraries and variable duration. The *Prinsendam*, belonging to *Holland America*, includes the navigation in Antarctica as part of a very long round the world itinerary. *The World* (only vessel that operates under the concept "residences at sea") visits Antarctica also as part of a voyage around the world. The *Silversea Cruise Line*, in the luxury segment, travels to Antarctica aboard the *Silver Explorer* that carries the same number of crew and passengers (117 as average during the 2014/2015 season).

As regards fleet renovation, the tour operator *Hurtigruten* has operated with a new vessel, *Fram*, since the 2007/2008 season. Other tour operators, like *Lindblad Expeditions*, have purchased vessels, such as the *National Geographic Explorer*

---

[18]Former *Marina Svetaeva*.

[19]Former *Akademik Schuleykin*.

[20]Former *Clipper Adventurer*, her original name was *Alla Tarasova*.

Fig. 5.6 Percentage of participation of tourist flows through Ushuaia from the 1957/1958 to 2014/2015 Austral summer seasons. *Source* Authors' own, based on Jensen and Daverio (2008), Vereda and Jensen (2014), Argentina (2015)

<table>
<tr><th>Decade</th><th>%</th><th>%</th><th>%</th></tr>
<tr><td>1950s</td><td>0.09</td><td rowspan="4">4.18</td><td rowspan="5">14.78</td></tr>
<tr><td>1960s</td><td>0.27</td></tr>
<tr><td>1970s</td><td>3.24</td></tr>
<tr><td>1980s</td><td>0.58</td></tr>
<tr><td>1990s</td><td>10.61</td><td rowspan="3">95.82</td></tr>
<tr><td>2000s</td><td>47.16</td><td rowspan="2">85.22</td></tr>
<tr><td>2010s</td><td>38.06</td></tr>
</table>

that was refitted and outfitted for expedition-style cruising. Another example corresponds to the French tour operator *Compagnie du Ponant*, which incorporated new technology in the vessels that they have acquired from 2010/2011 onwards: *Le Boreal* and *L'Austral* and *L'Soleal.*

Among the large vessels or "cruise only vessels" that call at the port of Ushuaia during the Antarctic voyage were: *Azamara Journey* (*Azamara Club Cruises*), *Celebrity Infinity* (*Celebrity Cruise Line*); *Prinsendam* and *Zaandam* (*Holland America Line*), all belonging to the premium segment.

During the first six summer seasons for the current decade 1,188 voyages took place, carrying 197,390 passengers, representing the 32.81 % of the total voyages and the 38.06 % of the total visitors (considering the whole period up to the 2014/2015 season).

Taking into account the whole period under analysis, 3,621 voyages took place, carrying 518,676 passengers. The 95.82 % of the total of visitors corresponds to the period between 1990 and 2014/2015 season, that is to say, to the moment the Antarctic product consolidated. The most significant concentration of tourist flows corresponds to the twenty-first century, representing the 85.22 % of the total (see Fig. 5.6).

Whether this trend will continue depends on several factors. One of the key issues that will influence the activity is related to the entry into force of the Polar Code for vessels that operate in polar waters together with the amendments to the SOLAS Convention and to the MARPOL Convention[21] that played an important role in giving the code a mandatory character from 1 January 2017 onwards. Also, in 2007 the IMO Assembly adopted Resolution A.999 "Guidelines on voyage planning for passenger ships operating in remote areas".

According to the flags of vessels that operated Antarctica throughout the last six seasons, it may be observed that most of them corresponded to non-Antarctic Treaty Party, with a participation that ranged from 56 % in the first season to 65 % in the last one. Most of them belong to Bahamas while the decrease of registries of

[21]Polar Code and the amendments to SOLAS Convention were adopted by IMO's Maritime Safety Committee (MSC) while the amendments to MARPOL were adopted by IMO's Marine Environment Protection Committee (MEPC).

	Number of vessels per season							
	Registries	2008/09	2009/10	2010/11	2011/12	2012/13	2013/14	2014/15
Antarctic Treaty Parties	France	1	1	1	2	2	2	2
	Germany	0	1	0	0	0	0	0
	Italy	0	1	0	0	0	0	0
	Netherlands	3	5	4	5	4	4	2
	Norway	1	1	1	1	1	1	1
	Russian Federation	9	6	5	3	3	3	3
	United Kingdom	0	0	0	0	0	1	0
	Bermuda (UK)	2	1	1	0	0	0	0
Non-Antarctic Treaty Parties	Antigua and Barbuda	1	1	1	1	1	1	0
	Bahamas	13	10	9	7	10	11	11
	Barbados	1	1	1	0	0	0	0
	Belize	0	1	1	0	0	0	0
	Cook Islands	1	1	0	0	0	0	0
	Cyprus	0	0	0	1	1	1	1
	Honduras	0	0	0	1	0	0	0
	Liberia	1	1	1	1	1	1	1
	Malta	1	2	2	1	1	2	1
	Marshall Islands	0	0	1	1	1	1	0
	Panama	2	1	0	0	0	0	0
	Union of Comoros	0	1	1	1	1	1	1

Fig. 5.7 Registries of vessels corresponding to the last six summer seasons. *Source* Authors' own, based on Vereda and Jensen (2014), Argentina (2015)

Antarctic Treaty Parties is due to the fewer Russian flagged vessels operating (see Fig. 5.7). Even though Antarctic tour operators are required to obtain an authorization from an Antarctic Treaty Party competitive authority, the flag state entails a very special importance as regards navigation.

With regard to visitors who went to Antarctica through the port of Ushuaia, passengers' nationalities were counted. In this respect, more than 70 nationalities were identified per season. Most of the passengers were from the United States with the 30 % approximately, followed by Germany, Australia, United Kingdom and Canada changing the order of the last four in the different seasons. It is interesting to point out that China passed from the sixth place in the 2009/2011 season to the third place in the last two seasons.

5.11 Final Remarks

Ushuaia has operated as a maritime tourism gateway since 1958 when Antarctic seaborne tourism started with the first commercial voyage aboard the *Les Eclaireurs*, organized by Argentina.

In the initial decades, Antarctic seaborne tourism was distributed between the private-sector initiatives and those promoted by different government agencies. Besides, it shows an erratic behaviour since a very remarkable fluctuation in the number and frequency of voyages could be demonstrated. In this respect, the 1970s had a significant movement with regard to the number of voyages that were organized and passengers carried.

From the 1990s onwards, Antarctic seaborne tourism was organized by initiatives coming from the private sector. Several exogenous and endogenous factors highly contributed to this development. Ushuaia started playing a very important role as the most active Antarctic seaborne tourism gateway due to two main reasons, on the one hand, the development of airport and port infrastructure as well as a diversity of specialized services has contributed to facilitate operations of Antarctic cruise ships and, on the other, the proximity to the Antarctic Peninsula has allowed tour operators to extend their Antarctic operating schedules.

References

Antartur. (n.d). *Cruceros Antártida Argentina, Manual del Crucerista*. (Handbook, supplemental material).

Argentina. (2015). Report on Antarctic tourist flows and cruise ships operating in Ushuaia during the 2014/2015 Austral summer season. *IP 126, XXXVIII ATCM*. Sofia, Bulgaria, June 01–10, 2015.

Arguindeguy, P. (1972). *Apuntes sobre los buques de la Armada Argentina, 1810-1970* (Vols. 3–6). Buenos Aires: Departamento de Estudios Históricos Navales.

ATCM IV. (1966). Recommendation ATCM IV-27. *Final Report IV Antarctic Treaty Consultative Meeting*. Santiago, Chile, November 03–18, 1966.

ATCM XXXII. (2009). Measure 15 ATCM XXXII. *Final Report XXXII Antarctic Treaty Consultative Meeting*. Baltimore, United States, April 06–17, 2009.

Bertram, E., Muir, S., & Stonehouse, B. (2007). Gateway ports in the development of Antarctic tourism. In J. M. Snyder & B. Stonehouse (Eds.), *Prospects for Polar Tourism* (pp. 123–145). Oxon: CAB International.

Borten, S. (1974). 10th Year of Lindblad Tourist Expeditions in Antarctica 1966–1975. American Society of Polar Philatelists. [Booklet].

Burbridge, H. (1984, August 29). [Letter to the Director of the Antarctic Board, Ministry of Foreign Affairs and Worship, Argentina]. Centro de Documentación Antártica, Universidad Nacional de Tierra del Fuego (Letter Nº 045). Ushuaia.

Capdevila, R. (2001). *Antártida. Más allá del fin del mundo*. Ushuaia: Zagier y Urruty Ediciones.

Capdevila, R., & Comerci, S. (2013). *Los tiempos de la Antártida. Historia Antártica Argentina*. Ushuaia: Aguafuerte Ediciones.

Coli, C. A. (2003). *La Armada Argentina en la Antártida. Campañas navales antárticas 1960-1980*. Buenos Aires: Edivérn.

Costa Cruise Fleet. The history of the company. Retrieved from http://www.costacruisesasia.com/B2C/SharedResources/Corporate/History/pdf/STORIA_COMPLETA_E.pdf (Para 8).

Delfino, A., & Cia, S. A. (n.d.) 18º Crucero de Turismo a la Tierra del Fuego con escala en Mar del Plata - 26 de Enero de 1929. [Brochure]. Retrieved from http://www.histarmar.com.ar/LineasPaxaSA/41-HamburgSud-Crucero1929Delfino.htm

De Agostini, A. M. (n.d.). *Mis viajes a la Tierra del Fuego*. Milan: Ed. Prof. G. De Agostini.

De Souza Reilly. (1933). En las tierras desconocidas de la Patria. Hacia el misterio de las Islas Orcadas. *Caras y Caretas*. Buenos Aires, March 11th 1933.
Diario La Nación. *Cálidas impresiones de los primeros turistas antárticos*. Buenos Aires, February 13th 1966 (pp 9). [Newspaper].
Dirección de Turismo Territorial. (1986). *Reseña de Turismo en Tierra del Fuego*. Ushuaia: Imprenta de Gobierno.
Dirección Nacional del Antártico. (1971). *Publicación Antártida N° 1*. Buenos Aires: Author.
Dirección Nacional del Antártico. (1999). *Argentina en la Antártida* (Vol. 2). Buenos Aires: Author.
Dirección Provincial de Puertos. (2007). Anuario del Puerto de Ushuaia. *Informe del Puerto y el turismo en la ciudad más austral del mundo*. Ushuaia: Desarrollos Editoriales.
Genest, E. A. (2001). *Antártida Sudamericana. Aportes para su comprensión*. Buenos Aires: Dirección Nacional del Antártico.
Genest, E. (2005). Presencia Argentina en la Antártida. In A. Molinari (Ed.) *La Argentina en la Antártida* (pp. 13–23). Buenos Aires: Consejo Argentino para las Relaciones Internacionales.
Guyer, R. E. (2006). Circunstancias que llevaron a la negociación y adopción del Tratado Antártico en 1959. In A. D. Abruzza (Ed.), *A cien años de la presencia permanente e ininterrumpida de la Argentina en la Antártida* (pp. 36–40). Ushuaia: Legislatura de la Provincia de Tierra del Fuego, Antártida e Islas del Atlántico Sur y Asociación Argentina de Derecho Internacional.
Headland, R. (1994). Historical development of Antarctic tourism. *Annals of Tourism Research, 21*(2), 269–280.
IAATO. (1999). Overview of Antarctic Activities. *IP 98, XXIII ATCM*. Lima, Peru, May 24-June 04, 1999.
IAATO. (2000). Overview of Antarctic Tourism. *IP 33, SATCM*. The Hague, Netherlands, September 11–15, 2000.
IAATO. (2009). Update on Kaptian Khlebnikov in Antarctica's Weddell Sea. *Press release*. Retrieved from http://iaato.org/c/document_library/get_file?uuid=c43c0bfa-9ee7-4c76-933b-a50c66dee7ad&groupId=10157
Jensen, M., & Daverio, M. E. (2008). Evolution of the Antarctic cruise ship tourism from Ushuaia, Argentina: 50 years as a maritime gateway to the Antarctic continent. *Proceedings Tourism and Global Change in Polar Regions. An International Conference* (pp. 84–91). University of Oulu, 29 November- 02 December, Finland.
Jutronich, J. (1958). En el país blanco de los hombres barbudos. *Revista Vea y Lea, Suplemento N° 114*, February 20, 1958 (pp. 1–8). Buenos Aires.
Meyer, D. (2004). Tourism routes and gateways: key issues for the development of tourism routes and gateways and their potential for Pro-Poor Tourism (pp. 1–31). *Discussion Paper*. Sheffield: Overseas Development Institute (Sheffield Hallam University).
National Research Council (1993). Science and Stewardship in the Antarctic. National Academy Press, Washington D.C. Retrieved from http://www.nap.cdn/openbook/030904947/html
Nichols, R. (1959). El primer viaje de turismo a la Antártida. *Argentina Austral* N° 334, Year XXXI, August 1959, pp. 4–5. Buenos Aires.
Palazzi, R. (1993). *Antártida y Archipiélagos Subantárticos. Factores para su análisis* (Vol. 3, N° 182). Buenos Aires: Escuela Superior de Guerra Aérea.
Palazzi, R. (2005). *La Argentina del Extremo Sur. 1810-2004*. Buenos Aires: Dunken.
Pearce, D. G. (2001). An integrative framework for urban tourism research. *Annals of Tourism Research, 28*(4), 926–946.
Pierrou, E. (1981). *La Armada Argentina en la Antártida 1939-1959*. Sus campañas en buques y bases. Buenos Aires: Instituto de Publicaciones Navales.
PNA. (2000). Inbound and outbound movement of vessels at the port of Ushuaia. *Book of registries 1949-2000*. Ushuaia: Prefectura Naval Argentina.
PNA. (2010). Inbound and outbound movement of vessels at the port of Ushuaia. *Book of registries 2001-2010*. Ushuaia: Prefectura Naval Argentina.
Reich, R. J. (1980). The development of Antarctic tourism. *Polar Record, 20*(126), 203–214.

Rodríguez, H. (1997). *Buques de la Armada Argentina 1970-1996*. Instituto Nacional Browniano: Sus comandos y operaciones. Buenos Aires.

Vereda, M. (2004). *El desarrollo de Ushuaia como puerta de entrada del turismo antártico. Tendencias y competitividad*. Unpublished Master's Thesis. Universidad Internacional de Andalucía, Sevilla.

Vereda, M., & Jensen, M. (2014). *La configuración del espacio turístico Tierra del Fuego, Antártida e Islas del Atlántico Sur. El rol de Ushuaia como puerta de entrada a la Antártida*. (Final Report Research Project Nº 898). Comodoro Rivadavia: Secretaría Ciencia y Técnica UNPSJB.

Index

M. Schillat et al., *Tourism in Antarctica*,
SpringerBriefs in Geography, DOI 10.1007/978-3-319-39914-0

MIX
Papier aus verantwortungsvollen Quellen
Paper from responsible sources
FSC® C105338

If you have any concerns about our products,
you can contact us on
ProductSafety@springernature.com

In case Publisher is established outside the EU,
the EU authorized representative is:
Springer Nature Customer Service Center GmbH
Europaplatz 3, 69115 Heidelberg, Germany

Printed by Libri Plureos GmbH
in Hamburg, Germany